Where Does Violence Come From?

Bernhard Bogerts

Where Does Violence Come From?

A Multidimensional Approach to Its Causes and Manifestations

 Springer

Bernhard Bogerts
Salus-Institut
Magdeburg, Sachsen-Anhalt, Germany

ISBN 978-3-030-81791-6   ISBN 978-3-030-81792-3 (eBook)
https://doi.org/10.1007/978-3-030-81792-3

© The Editor(s) (if applicable) and The Author(s), under exclusive license to Springer Nature Switzerland AG 2021
This work is subject to copyright. All rights are solely and exclusively licensed by the Publisher, whether the whole or part of the material is concerned, specifically the rights of reprinting, reuse of illustrations, recitation, broadcasting, reproduction on microfilms or in any other physical way, and transmission or information storage and retrieval, electronic adaptation, computer software, or by similar or dissimilar methodology now known or hereafter developed.
The use of general descriptive names, registered names, trademarks, service marks, etc. in this publication does not imply, even in the absence of a specific statement, that such names are exempt from the relevant protective laws and regulations and therefore free for general use.
The publisher, the authors and the editors are safe to assume that the advice and information in this book are believed to be true and accurate at the date of publication. Neither the publisher nor the authors or the editors give a warranty, expressed or implied, with respect to the material contained herein or for any errors or omissions that may have been made. The publisher remains neutral with regard to jurisdictional claims in published maps and institutional affiliations.

Cover illustration: @Ivy Close Images / Alamy Stock Photo

This Springer imprint is published by the registered company Springer Nature Switzerland AG
The registered company address is: Gewerbestrasse 11, 6330 Cham, Switzerland

Preface

For many people, violence is an incomprehensible phenomenon that occurs in a multitude of forms and can affect all areas of life. A prerequisite for curbing violence is knowledge of its manifold manifestations and causes. Due to the importance of this topic, it is not surprising that several excellent and quite comprehensive books have been published recently. However, these books illuminate the problem of violence from the perspective of various different disciplines. These include the influential book by Pinker [1], which focuses on evolutionary, historical and psychological aspects, the books by Raine [2], Sapolsky [3] and Haller [4] with an emphasis on neurobiological views, the books by Dwyer and Micale [5], as well as by Gerlach [6], from a historical perspective, the book by Lee [7] focusing on psychological, social and political approaches, and the book by Armstrong on religion and violence [8], to name but a few. In the available interdisciplinary literature on the subject [9] there are usually subject-specific individual contributions by various authors, while integrative approaches are neglected.

The motivation for writing this book was to offer an updated multidimensional view of the phenomenon of violence, summarizing the various sub-disciplines, in a manageable and generally understandable form, taking into account the state of literature worldwide. In order to achieve this goal, the complexity of the subject matter often made it necessary to simplify the presentation of neuroscientific, genetic, psychological and social science issues. References to scientific details and further publications can be found in the extensive bibliography.

The author of this book is a psychiatrist and neuroscientist, so his core competencies lie in the clinical and neurobiological fields. However, the complex set of conditions of the many facets of violence can only be understood through an integrative view of neurobiological, psychological, psychopathological and sociological points of view. The inclusion of sociological knowledge in this book was accomplished with the help of Christian Steinmetz, MA, research associate at the Salus Institute.

The book is structured in such a way that first the extent and type of occurrence of various manifestations of physical violence are presented. Subsequently, the principles of evolutionary biological, genetic and neuroscientific causes are explained in language understandable for lay people. This is followed by a summary of theories and findings from psychology, psychiatry and the social sciences with special emphasis on mental disorders, hedonistic and collective violence as well as the sensitive topic of religion and violence. The chapters on brain pathology, amok, terror and hedonistic violence are presented with striking examples.

A summary of several different scientific fields in such a far-reaching and complex subject area by a single author will imply that some subject areas cannot be presented in their entire complexity and that experts of partial disciplines will know how to present additional information. Intensifying a cross-disciplinary dialog to explore the causes of violence is a major concern of this book, thereby setting the stage for improving prevention.

Magdeburg, Germany
May 2021

Bernhard Bogerts

References

1. Pinker S (2011) The better angels of our nature: why violence has declined. Viking Books Adult
2. Raine A (2013) The anatomy of violence—the biological roots of crime. Penguin Books, London
3. Sapolsky RM (2017) The biology of humans at our best and worst. Penguin Press, New York
4. Haller J (2020) Neurobiopsychological perspectives on aggression and violence. Springer Berlin Heidelberg
5. Dwyer P, Micale MS (2020) On violence in history. 1. Berghahn Books, New York
6. Gerlach C (2010) Extremely violent societies: mass violence in the twentieth-century-world—Chapter 4. Cambridge University Press
7. Lee BX (2019) Violence—an interdisciplinary appoach to causes, consequences, and cures. Wiley Blackwell
8. Armstrong K (2014) Fields of blood: religion and the history of violence. 1. Random House, London
9. Van Hasselt VB, Bourke ML (eds) (2017) Handbook of behavioral criminology. Springer International Publishing, Cham. https://doi.org/10.1007/978-3-319-61625-4

Acknowledgments

I am indebted to Christian Steinmetz, research assistant at the Salus Institute in Magdeburg, for his help with socio-scientific content and illustrations. My special thanks go to Thomas Kluger, who contributed helpful comments on all chapters of the book in many friendly discussions, and also to Prof. Dr Anne-Maria Möller-Leimkühler for her valuable suggestions on the overall concept of the book. I thank my sisters for critically checking the text for lay comprehensibility and readability. Last but not least, I have to thank Mr. Fietz-Mahlow, Managing Director of Salus-Altmark Holding, for creating excellent conditions for writing this book.

Where Does Violence Come From?

Why do people do such things? This is often the first question that arises when we witness violence in the in real life or read about it in the media. This book provides comprehensive answers: rather than explaining the causes of violence from the limited perspective of a single discipline, it combines explanatory approaches from criminology, sociology, psychology, psychiatry, brain research, genetics, pedagogy, historical sciences and justice into a big, exciting and comprehensible picture in an entertaining way and according to the current state-of-the-art science(s). And always close to case studies that show us the frightening diversity of human violence: acts of violence by individual perpetrators, violence between groups, riots by gangs and hooligans, violent ethnic and religious conflicts, extreme violence in the form of amok and terror, and armed conflicts, pogroms and genocide. Last but not least, the knowledge gained from this book can help answer another big question: how can violence be contained or even prevented?

- How and where does violence originate in our brain?
- Why has a tendency towards violence become established as part of our behavioral repertoire in the development of humankind?
- What influences on personality development can lead to violent characters? How often is violence the product of a pathological psyche? Do genes play a role?
- Which social constellations contribute?
- What are the causes of rampage and terror?
- What is known about the relationship between religion and violence?

Contents

1	**Introduction**	1
2	**Manifestations of Violence**	3
	Classification into Individual or Group Violence	3
	Classification by Cause and Motivation	4
	References	5
3	**Incidence, Frequency and Consequences of Violence**	7
	Dimensions of Violence in Global Comparison	7
	Partner Violence	9
	Violence Towards Children	10
	Long-Term Psychological and Economic Consequences of Violence	11
	Current Situation in Historical Comparison	13
	References	14
4	**Why is the Tendency to Violence a Human Trait?**	17
	Aggression and Violence as a Result of Human Evolution: Phylogenetic Causes	17
	Changing Character Traits by Selective Breeding	18
	Parallel Development of Aggression and Sympathy During Evolution	19
	Why did Prehumans and Early Humans Disappear?	20
	Decrease in Violence with Increasing Civilization?	23
	Phylogenesis of Prosocial Behavior	24
	References	26
5	**Heritability of Aggressive Behavior**	29
	Significance of Genes	29
	The Interplay of Genes and Environment: Epigenetics	30
	How Strong is the Influence of Genes? Twin and Family Studies	31
	Which Genes Play a Role?	33
	What Do Genes Do in the Brain?	33

	Can Gene Analyses Predict Dangerous Behavior?	34
	Genes and Prosocial Traits	34
	Genes and the Future of Our Behavior	35
	References	35
6	**Neurobiology of Violence**	**37**
	Evidence of Aggression Centers in the Brain	37
	Regulation and Control of the Aggression Centers in the Brain	41
	Phylogenetic Tripartition of Brain Structure and Function: Concept of the Limbic System	42
	Stages of Information Flow Through the Brain	43
	Connection Between Violence and Reward Centers	44
	Neurobiology of Prosocial Behavior	45
	Neurobiological Correlates of Ethics and Morals	47
	Brain Activity in Empathy and Compassion	47
	References	48
7	**Brain Pathology in Violent Offenders**	**51**
	Examining the Brain Using Imaging Techniques	51
	Causes of Brain Structural and Functional Deficits	53
	Historical Cases—Prominent Examples	53
	Phineas Gage	53
	Head Schoolmaster Ernst Wagner	54
	Charles Whitman	54
	Ulrike Meinhof	56
	Brain Pathology of Imprisoned Violent Offenders	57
	References	58
8	**The Role of Hormones and Messenger Substances in the Brain**	**61**
	Testosterone	61
	Oxytocin	62
	Stress Hormone Cortisol	62
	Serotonin	63
	References	64
9	**Gender Difference in the Propensity for Violence**	**67**
	Phylogenetic Causes	67
	Brain Biological Correlates of Sex Difference	68
	References	68
10	**Mental Disorders and Violence**	**71**
	General Risk of Violence in Mental Disorders	71
	Schizophrenic and Psychotic Diseases	72
	Depressive Disorders	74
	Bipolar Disorders	74
	Attention Deficit Hyperactivity Disorder (ADHD)	74
	Brain Injuries and Tissue Defects	75

	Post-traumatic Stress Disorder (PTSD)	75
	Borderline Personality Disorder	76
	Dissocial/Antisocial Personality Disorders	76
	Psychopathy	77
	Narcissistic and Histrionic Personality Disorders	77
	Paranoid Personality Disorder: Fanatics	78
	Pathological Irascibility, Rage Syndrome, Cholerics	78
	Risk of Violence Due to Personality Disorders	79
	References	79
11	**Alcohol, Drugs and Violence**	83
	Addiction as Cause and Consequence of Violence	83
	Frequency of Violence Under the Influence of Alcohol	84
	Effect of Alcohol on the Brain	85
	Effects of Drugs	85
	Drug Terror	86
	References	87
12	**Psychology of Violence**	89
	Historical Attempts to Explain Violence	89
	The Drive Theories of Freud and Lorenz	92
	Frustration Theory and Learning Theory	92
	Violence: A Product of Civilization?	93
	The Banality of Evil	93
	New Psychological Theories of Aggression	95
	The Dark Tetrad of Personality	97
	References	98
13	**Violence as an End in Itself and Lust Gain**	101
	Current and Historical Examples	101
	Torture and Sadism	103
	Sadistic Serial Killers	104
	Revenge	104
	Collective Violence as a State of Ecstasy	105
	Hedonistic Violence as a Relic of Phylogenesis	106
	Brain Biological Correlates of Hedonistic Aggression	108
	References	110
14	**Social Causes of Violence**	113
	Historical and Geographical Variations in the Frequency of Violence	113
	The Importance of the State Monopoly on the Use of Force	114
	The Downsides of the State Monopoly on the Use of Force	117
	Police Violence	118
	Economic Conditions and Violence	118
	Societal Attitudes Toward Violence	119
	Anomie and Disintegration as Causes of Violence	120

	Interaction of Social, Biographical and Neurobiological Conditions	122
	References	123
15	**Violence in Children and Adolescents: Early Risk Factors**	127
	Neurobiology of the Maturing Brain	127
	Occurrence and Frequency	128
	Causes of Aggression in Childhood and Adolescence	129
	Early Traumatization	129
	Hereditary Factors	129
	Mental Disorders: ADHD	130
	Brain Development Disorders: Toxic Influences	130
	New Media and the Risk of Violence Among Young People	131
	Predictability of Future Violence in Children and Adolescents?	131
	References	133
16	**Rampage and School Shooting**	137
	Difference Between Rampage and Terror	137
	Deadliest Rampage Events of the Last 20 years	137
	Most Momentous Rampage Killings Occur in the US	140
	Deadliest School Shootings Worldwide	140
	Who Becomes a Mass Murderer?	143
	Studies of Surviving Spree Killers	143
	Further Research Projects on the Psychology of Rampage Killers	144
	School Shooting Preventive Measures	145
	Early Warning Symptoms: "Leaking"	146
	Warning Symptoms in Adult Spree Killers	147
	Which Brain Functions Are Damaged in Spree Killers?	147
	Rampages in the Prodromal Stages of Schizophrenic Disorders	149
	Future Risk of Rampages and School Shootings	150
	References	150
17	**Terror**	155
	What is Terror?	155
	Historical Background and Current Developments	156
	The Increasing Importance of the Internet	157
	Is Terrorism on the Rise?	158
	Who Becomes a Terrorist?	159
	Psychology and Sociology of Group Terrorists	159
	Left-Wing Terrorism	159
	Right-Wing Terrorism	160
	Right-Wing Terrorism as a Predominantly Male Phenomenon	161
	Islamist Terrorism	161
	Special Features of Salafist Terrorism	163
	Common Characteristics of Terrorist Groups	163
	Characteristics of Lone-Wolf Terrorists	165

	Most Momentous Lone-Wolf Attacks in the Last Decade	166
	How to Identify Lone Wolves?	167
	Brain Structure and Brain Function of Terrorists	167
	Coincidence of Personality and Environment in Terrorists	168
	References	169
18	**Collective Violence, Xenophobia, Pogroms, Genocide**	173
	Collective Violence as a Legacy of Evolution	173
	Similarities Between Humans and Animals	173
	Historical Dimensions of Collective Violence	175
	Risk Factors for Wars and Genocides	177
	Sociological Studies on the Emergence of Group Hatred and Violence	178
	Social Identity	178
	Established and Outsiders	180
	Realistic Group Conflicts	180
	Felt Threat	181
	Group-Based Misanthropy: Right-Wing Extremism	181
	Group Violence as a Male Domain	182
	Suspension of Inhibition Mechanisms: Behavior in War	182
	Disinhibition as a Phenomenon of Mass Psychology	184
	Brain Biological Correlates of Group Aggression and Racism	184
	Getting to Know Each Other Against Prejudice	185
	References	186
19	**Sexual Violence**	189
	Definition	189
	Frequency	190
	Types of Perpetrators: Risk Factors	191
	War and Sexual Violence	192
	Phylogenetic Aspects	192
	References	193
20	**Religion and Violence**	195
	Common Characteristics of the Great Religions	195
	Violence in the Name of the Religions	196
	Islam	196
	Christianity	197
	Judaism	200
	Buddhism and Hinduism	201
	Sects	202
	Psychological and Sociological Explanations	203
	Neuroscientific Explanatory Models for the Relationship Between Religion and Violence	204
	Religious Phenomena and Violence in Brain Diseases	205

	Limits of Cognition and Knowledge	206
	References	207
21	**Conclusions for the Prediction and Prevention of Violence**	209
	Limits to the Predictability of Individual Violence	209
	Predictability of Collective Violence	211
	Phylogenetic Disposition to Individual and Collective Violence Remains Unchanged	212
	Current Situation for the Prevention of Violence	212
	Prevention Principles and Projects	213
	Successful Evidence-Based Projects	214
	Final Remarks	215
	References	216
Index		219

About the Author

Prof. Bernhard Bogerts, MD is a neuroscientist and psychiatrist. From 1994 to 2015, he was clinical director and full professor of psychiatry at the University of Magdeburg, Germany. Since his retirement, he has been the director of the Salus Institute in Magdeburg, whose scientific focus is research into the causes of violence. He has received several awards for his research on the brain-biological basis of mental disorders. He also became known for his work on psychological and brain pathological findings in violent offenders.

Chapter 1
Introduction

Human coexistence is mainly characterized by peaceful cooperation; interpersonal harmony determines our lives much more often than dissonance. Violence, however, sometimes appears as spontaneous or planned behavior of single persons or whole groups, but sometimes also—and there are numerous examples of this at present and in history—as a looming mass phenomenon that can eventually take on apocalyptic proportions.

Why does violence exist in its various forms at all? Violent acts of individual perpetrators, violence between groups, vandalism and riots by gangs and hooligans, violent ethnic and religious conflicts, extreme violence ranging from rampage or terror to armed conflicts and genocide. How and where does violence arise in our brain? Why has the propensity for violence established itself in the development of mankind as a not inconsiderable part of our behavioral repertoire? What influences on personality development can lead to violent characters? How often is violence the product of a mental disorder? Do genes play a role? Which social constellations contribute to this?

This book offers an integrative view of the phenomenon of violence, which various disciplines such as criminology, sociology, psychology, brain research, genetics, pedagogy, history and justice otherwise try to explain from different perspectives, often without further consideration of the findings of neighboring fields of science. In particular, the social sciences, which currently dominate the formation of opinion on this topic, will be supplemented by neuroscientific, phylogenetic, psychological and psychiatric aspects.

Chapter 2
Manifestations of Violence

Aggression and resulting violence are complex phenomena with multiple interrelated causes. Violence occurs not only in its physical form with the aim of physically damaging, subjugating, eliminating or annihilating others. More common are practices ranging from psychological aggression in the form of intrigue, stalking, bullying, cyber-bullying, defamation, exclusion up to psychological terror with all its variants, the inventiveness of which sometimes seems unlimited. No less significant is so-called structural violence, which refers to the oppression and exploitation of entire groups of people.

Due to the multidimensional character of violence, which is often the subject of controversial discussions between social scientists, psychologists and neurobiologists, it is not surprising that there are different views on the causes, definitions, classification criteria—and prevention—of violence.

This book is mainly devoted to physical violence, which aims to physically damage, subjugate, repress or kill others. The many forms of psychological violence and structural violence, which can have consequences as disastrous as direct physical violence and often precede it, are not the focus of this book.

The following forms of physical violence can be distinguished.

Classification into Individual or Group Violence

(a) **Individual violence**, in which an individual person becomes violent against one or more other persons, for example, in the form of assault, deprivation of liberty, rape, murder, manslaughter or even rampage.
(b) **Collective violence**, where a group of people attacks another group or individuals, ranging from riots and brawls by hooligans, clashes between gangs, tribes, radicalized political groups and religious communities to pogroms, wars and genocide. Collective violence also includes expulsions, deportations and

resettlements, which, even without the use of direct physical violence, have often resulted in mass deaths of those affected.

The protagonists of collective violence—similar to terrorists—usually invoke ideologies justifying their actions.

(c) **State violence** in the form of a monopoly on the use of force to maintain and secure a political or social system, to enforce legal norms and to protect citizens. Numerous examples from history show, however, that state violence is not only intended to maintain order and serve security-guaranteeing goals but can also, depending on the type of political system and ideologies sanctioning violence, take on immense forms of state terror.

Classification by Cause and Motivation

Irrespective of the number of people involved and the type of exertion, violence can be classified according to cause or motive.

(a) **Reactive violence**, which is triggered by provocation or threat and aims to eliminate it. Reactive violence in a broader sense also includes revenge, that is, the urge to repay the damage in equal measure. Reactive perpetrators of violence often include the delinquents described by the judiciary and forensic psychiatry as perpetrators of affect.
(b) **Proactive violence**, which is deliberate violence planned in advance to gain personal advantage by harming others. The aims are the exertion of power, the pursuit of dominance, enrichment, greed, subjugation, expulsion or elimination of others without prior provocation by the victims. This includes predatory and exploitative violence, sexual violence, violence for the purpose of dominance or to maintain power, but also hedonistic violence, which is exerted for its own sake because it is fun and thus serves to increase pleasure, as well as sadism and torture.
(c) **Revenge and retribution** as a combination of reactive and proactive violence. In the transitional area between reactive and proactive violence there is also violent rebellion against actual or perceived oppression and exploitation.
(d) **Violence as a result of a mental disorder or brain damage**. This includes delusional symptoms in psychotic disorders, mood disorders, pathological fanaticism and damage of certain violence-controlling areas of brain tissue.

This division of violence is not to be understood as a drawer-like separation of the different forms listed here; often, seamless transitions or combinations are encountered. Reactive violence can be combined with planned or disease-related violence, individual with collective violence.

Galtung [1] introduced the terms "**structural violence**" and "**cultural violence**" into the discussion. Distinct from direct personal violence, this refers to repression by political and social structures, cultural systems or ideologies, which prevent people from realizing their own potential [1] without concrete actors exercising physical violence being identifiable. The consequences of oppression, exploitation, exclusion, extreme income inequality, ramshackle legal systems, modern slavery, associated poverty, inadequate medical care, and lack of food and other essentials can undoubtedly be even more disastrous than the consequences of direct physical violence [2].

The term "structural violence", however, was criticized as being blurred and open to arbitrary interpretation, since almost all social injustices could be described in this way [3–5]. In addition, it was objected that any form of direct physical personal exercise of violence in the experience of those involved is completely incomparable with what is referred to as structural violence. Also, it is always concrete social actors or groups of people who, motivated by the desire for power, dominance or possession, make use of certain structures, political systems or ideologies in order to expand their own possibilities and restrict those of others [3]. There are always persons behind structures. Thus, the concepts of structural and cultural violence always include an—albeit indirect—form of personal oppression of others.

References

1. Galtung J (1969) Violence, peace, and peace research. J Peace Res 6(3):167–191. https://doi.org/10.1177/002234336900600301
2. Lee BX (2019) Violence—an interdisciplinary approach to causes, consequences, and cures. Wiley Blackwell
3. Riekenberg M (2008) Auf dem Holzweg? Über Johan Galtungs Begriff der "strukturellen Gewalt" (On the wrong track? On Johan Galtung's concept of "Structural Violence"). In: Zeithistorische Forschungen, pp 172–177
4. Lawler P (1989) A question of values: a critique of Galtung's peace research. Interdiscip Peace Res 1(2):27–55. https://doi.org/10.1080/14781158908412711
5. Dilts A, Winter Y, Biebricher T, Johnson EV, Vázquez-Arroyo AY, Cocks J (2012) Revisiting Johan Galtung's concept of structural violence. New Polit Sci 34(2):e191–e227. https://doi.org/10.1080/07393148.2012.714959

Chapter 3
Incidence, Frequency and Consequences of Violence

The frequency and extent of individual and collective violence is subject to considerable fluctuations depending on the world region as well as the historical and social situation. A short description including some statistical data on the intensity of the worldwide problem of violence is presented in the following.

Dimensions of Violence in Global Comparison

In Europe, we currently live in a relatively safe region of the world. The number of homicides in Central Europe is, at about 0.8 per 100,000 inhabitants per year, [1, 2] at a statistically low level compared to four to ten times higher rates in several Eastern European regions and the United States and up to 40 times higher rates of murder and homicide in some countries in Africa and Latin America [2, 3]. The relatively high homicide rate in the United States of about 6 per 100,000 is due to the general availability of firearms. More than two-thirds of all homicide victims in the US are shot [4]. In addition, there is an even higher number of suicides by firearms (2018: 7/100,000) [4]. An overview of the frequency of homicides in different regions of the world is given in Fig. 3.1.

The risk of being a victim of homicide varies by more than a 100-fold across world regions. In 2017, the lowest homicide rates were in Singapore and Japan (0.2 and 0.3 per 100,000 population/year, respectively), and the highest were in Central America: El Salvador 62 per 100,000 population, Venezuela 57, Honduras 41, followed by South Africa 34. Despite a significant increase of murder and manslaughter in several Central American and African countries, the global average rate of homicides has decreased by about 20% in the last 15 years from 7.80/100,000 in 2006 to ca. 6.4/100,000 in 2018 (see Figs. 3.1 and 3.2) [2, 5].

Globally, violence is one of the leading causes of death among younger and middle-aged adults. Depending on the institution collecting the data, fatality figures

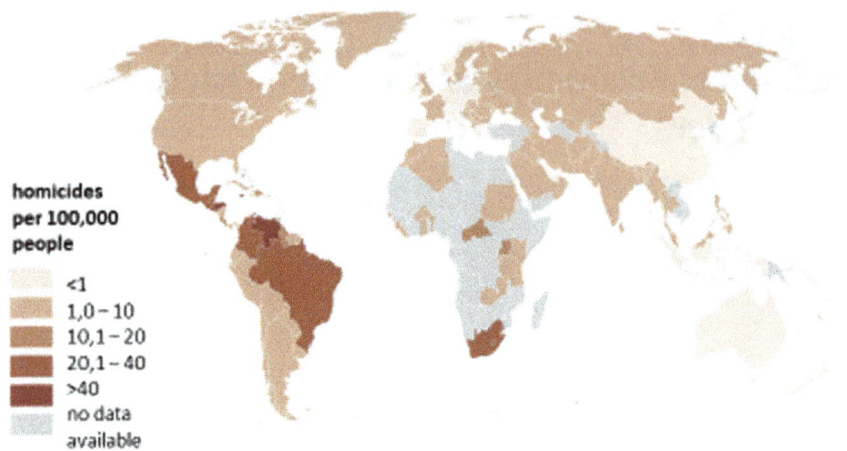

Fig. 3.1 Global homicide rates (per 100,000 population) in 2017. *Source* UNODC Global Study on Homicide [2]

Fig. 3.2 Course of global homicide rates since 2007. *Source* Global Peace Index 2020 [5]

for 2015–2017 ranged from 460,000 to 600,000 each year [2, 6, 7]. Of these, two-thirds were victims of individual violence, and one-third of collective violence [8, 9]. Young men aged 14–29 were five times more likely to die than other age groups.

Criminal homicide causes more deaths than armed conflicts and terrorism combined. According to the United Nations Office on Drugs and Crime (UNODC), in 2017, the global death toll from homicide was 464,000, significantly higher than that from armed conflicts (89,000) and terrorist attacks (26,000) [2]. Organized crime alone killed approximately as many people around the globe in 2017 as all armed conflicts combined.

Forty percent of global homicides occur among children, adolescents and young men, with an estimated 200,000 deaths in this age group, according to the WHO (2015) report [10]. Nearly half of boys and about a quarter of girls aged 13–15 report having been involved in acts of physical violence either as perpetrators or victims.

Every year, 9 out of 10 violent crimes, such as murder, manslaughter, rape, grievous bodily harm and damage to property, are carried out by men [1]. However, in the statistics of non-violent crimes, such as fraud, slander and theft, both genders

Fig. 3.3 Suspects by age group and gender per 100,000 persons of the same age and gender. *Source* Constance Crime Development Inventory [12]. *Data source* German Police Crime Statistics 2017

come quite close (see Fig. 3.3). In addition to male gender, age is a second risk factor for violent offenses; the 15–25 age group is disproportionately represented; as age increases, the incidence of such offenses decreases substantially [11]. Thirty times more men than women are imprisoned for assault. Acceptance of violence, right-wing extremist orientations and the use of violence by hooligans are the domain of male youths.

Partner Violence

Violence in partnerships is multifaceted: physical violence, manslaughter, murder, sexual violence, threats, psychological violence in the form of humiliation, coercion and isolation, kidnapping and stalking. The perpetrators are predominantly those characters with an already low empathy and an increased tendency to violence with an absolute claim to power, which they can live out at home with the least inhibition towards the physically weaker female partner. They hardly ever blame themselves for their actions, but regularly blame the victim [13].

Globally, the proportion of women who experience intimate partner violence or sexual violence is estimated at 35% [14, 15]. 80% of all murders of women are perpetrated by intimate partners; this amounted worldwide to 50,000 cases in 2017 [2]. The figures are only an approximation of the actual situation, as no reliable data are

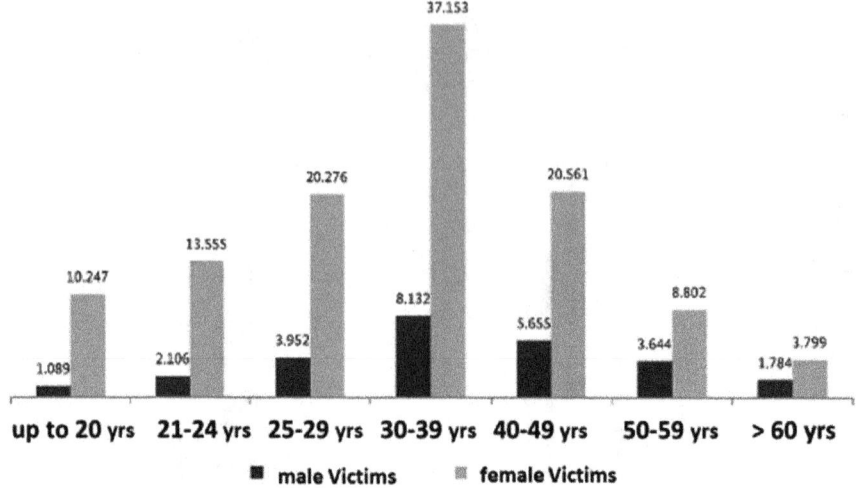

Fig. 3.4 Victims of intimate partner violence by gender and age group. *Source* Bundeskriminalamt (German Federal Criminal Police Office) 2019 [16]

available from many countries; this is particularly true for the Eastern Mediterranean region, Arab countries and African states. Violence against women rarely occurs in public, but mainly in private, which is why many such acts escape public attention.

As a representative example for Western Europe, figures for Germany show that almost a quarter of the cases of murder and manslaughter occurred in a partnership; men were seven times more likely to be perpetrators than women. More than half of the victims lived in the same household as the perpetrator. 80% of the victims were female, 20% male (see Fig. 3.4). Violence within partnerships accounted for about 17% of all violent acts [16]. The number of unreported cases might be significantly higher; only about 20% of those affected file a report.

Violence at home is experienced as particularly traumatizing because the place of domestic security is no longer available.

Violence Towards Children

According to UNICEF's 2017 report, [17] three-quarters of two- to four-year-olds worldwide (nearly 300 million) and more than half of six- to ten-year-olds suffer violence in the form of physical punishment or psychological aggression at the hands of parents or caregivers. However, the actual numbers are difficult to estimate because the child victims hardly have the means of reporting it themselves.

In 2017, 21,000 children under the age of 14 died a violent death worldwide [18]. Every seven minutes, somewhere in the world, an adolescent is killed by an act of violence [17]. Male adolescents are particularly at risk; they are four times more likely than girls to be victims of homicide. Homicide rates for adolescent boys are particularly high in Central and South America, with mortality rates up to nearly 100 deaths per 100,000 people in Venezuela and 60–70/100,000 in Columbia, El Salvador, Honduras and Brazil [17]. Girls, on the other hand, are more at risk from sexual violence.

Corporal punishment at school is meanwhile prohibited in the majority of countries worldwide (with the exception of some African and Arab states), but 90% of children under five live in countries where corporal punishment at home is not prohibited by law. Worldwide three in four young children are still regularly subjected to violent discipline by their caregivers [17, 19].

Nor is child maltreatment uncommon in Western Europe. In Germany, for example, according to police crime statistics, in 2017 more than 4000 children were victims of violence, with approximately 2000 male and 1500 female adult suspects [20]. Violent acts ended fatally in 140 children. The number of victims has increased slightly over the last 15 years. The acts are most often consequences of excessive demands and frustration when a parent is emotionally unstable. Not listed in these statistics are the cases of sexual abuse of children, which are associated with a high number of unreported cases.

Children who experience violence are often traumatized for the rest of their lives; they have a higher risk of emotional disorders in adulthood, developing addictions or committing violent acts themselves, and are less likely to achieve higher school qualifications [21].

Long-Term Psychological and Economic Consequences of Violence

The global costs of violence and its consequences have been estimated at a total of US\$9–10 trillion (10^{12}) per year, or about 11 percent of global economic output [22]. By comparison, military spending is much lower [23] (see Fig. 3.5).

By far the highest follow-up costs of violence worldwide are caused by domestic violence (violence by the partner and child abuse). These costs were estimated to be six to seven times higher than non-domestic violence and more than 50 times higher than those caused by group violence, including terror [22, 23].

In contrast to civil war, terror and murder, violence against partner and children—although quantitatively much more important—is not in the forefront of media coverage or at the center of political attention, probably because domestic violence takes place in the private sphere and seems to be almost a daily occurrence.

More serious than the economic consequences are the long-term psychological effects of violence, such as post-traumatic stress disorders and persistent personality changes after extreme stress in the form of depression, apathy, anxiety and fright. Children who experience violence in the first years of their lives are more vulnerable to later mental illness, school failure, alcohol and drug addiction, but also to

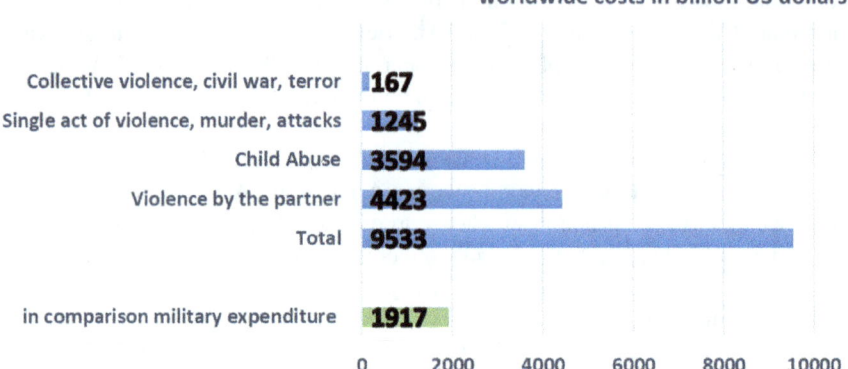

Fig. 3.5 Worldwide costs of violence compared with military spending. Violence in its various forms caused worldwide follow-up costs of approximately US$ 9.5 trillion (=9500 billion). This corresponds to 11.16% of global gross domestic product (latest data from 2013), [22] compared to military spending of US$1.9 trillion (2019) [24]. Own diagram. Data source see Ref. [22]

physical illnesses such as diabetes and cardiovascular diseases. In addition, children who grow up under violent conditions have a high risk of becoming perpetrators of violence themselves later or continuing to be victims of violence. Poverty, inequality and exclusion, and the feeling of being treated unfairly in turn increase the risk of violence for those affected. Increased violence in certain regions of the world in turn reduces economic investment and thus leads to further poverty. This again encourages violence, resulting in a vicious circle.

According to United Nations estimates, in 2018 alone, around 70 million people were fleeing violent conflicts or human rights violations in their countries of origin. Forty-one million of them remained refugees and displaced persons in their own country, 26 million were distributed among other countries as refugees (see Fig. 3.6). Among these refugees were approximately 100,000 children without parents or other caregivers. The reasons for these refugee flows in recent years include the conflicts in Syria, Somalia, Sudan, Colombia, Central Africa and Iraq [5, 25].

Fig. 3.6 Dramatic increase of refugees and displaced people, 2009–2019. *Source* Global Peace Index 2020 [5]

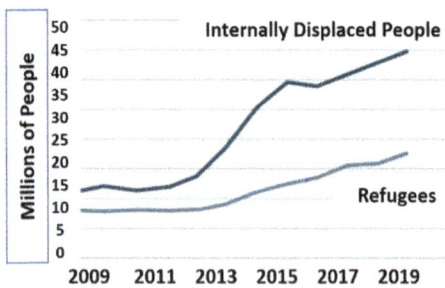

Current Situation in Historical Comparison

Not only compared to other regions of the world, but also from a historical perspective, in high-income countries we are now living in a very peaceful time; the level of violence has never been lower than it is today.

Faced with almost daily coverage of violence, wars and terror around the world, it is easy to lose sight of the fact that since the end of the Second World War, Europe has experienced the longest period of peace since the beginning of history. At least in Europe (apart from the wars and genocide in the former Yugoslavia between 1991 and 1999 and the fighting in Eastern Ukraine since 2015) and in Japan, the lessons learned from the catastrophic consequences of fascism, the Nazi era and Stalinism have been lasting. However, this is not true for all parts of the world. Moreover, the question arises to what extent the resurgence of nationalism and populism in Europe and America in recent years, combined with growing xenophobia, is sowing the seeds for a development reminiscent of the national state thinking at the beginning of the last century and its consequences.

In order to give a historical appreciation of our present situation, it should also be mentioned that between the late Middle Ages and the twentieth century there was a more than 30-fold decline in the murder rate in relation to population numbers in Western European states [26, 27] (see Fig. 3.7). This development over the last four centuries can be attributed to the increasing outlawing of violence as a result of the rise of humanism, the end of the medieval feudal system, the Renaissance, the Enlightenment, increasing education, better living conditions and, last but not least, the takeover of the monopoly of violence by the state [26].

However, this positive development has been limited to the Western world. Murder, manslaughter, expulsion and extermination of the indigenous population by Europeans in the colonies they occupied have continued with unchanged intensity in recent centuries [28, 29]. Even today, the rate of violent acts, murder and manslaughter in some countries of the so-called Third World (see Fig. 3.1) is comparable to the situation in Europe in the Middle Ages.

Despite this relatively peaceful world in historical and global comparison, at least in the more highly developed industrial nations, individual and collective violence remains humanity's biggest problem worldwide, alongside hunger (which is often a consequence of oppression and violence).

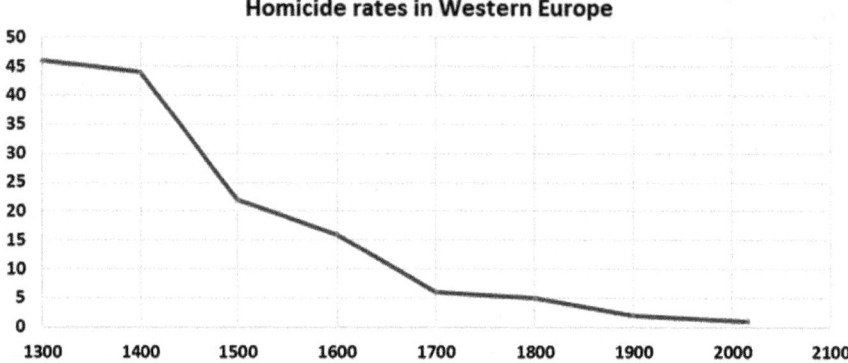

Fig. 3.7 Murder rates from the thirteenth century to 2010, selected countries or regional groups in Western Europe. Mean values from seven European countries: Netherlands, Belgium, Scandinavia, Italy, Germany, Switzerland, England. Number of murders per 100,000 persons/year. *Source* Our World in Data: Eisner (2003) and IHME, Global Burden of Disease (2017) [27]. Own diagram

References

1. Bundeskriminalamt. Polizeiliche Kriminalstatistik (2019) (Police crime statistics 2019, German Federal Criminal Police Office). https://www.bmi.bund.de/SharedDocs/downloads/DE/publikationen/themen/sicherheit/pks-2019.pdf?__blob=publicationFile&v=10. Published 2019
2. UNODC United Nations Office on Drugs and Crime (2019) Global study on homicide. Vienna. https://www.unodc.org/unodc/en/data-and-analysis/global-study-on-homicide.html
3. WHO Global Status Report on Violence Prevention. Geneva. https://www.who.int/violence_injury_prevention/violence/status_report/2014/en/. Published 2014. Accessed 22 September 2020
4. Goldstick JE, Carter PM, Cunningham RM (2020) Current epidemiological trends in firearm mortality in the United States. JAMA Psychiat. https://doi.org/10.1001/jamapsychiatry.2020.2986
5. Institute for Economics & Peace. GLOBAL PEACE INDEX 2020: measuring peace in a complex world. https://www.visionofhumanity.org/maps/. Published 2020. Accessed 2 December 2020
6. WHO. 10 facts about violence prevention. http://www.who.int/features/factfiles/violence/en/. Published 2017
7. Bellis M, Hardcastle K, Hughes K, Wood S, Nurse J (2017) Preventing violence, promoting peace—a policy toolkit for preventing interpersonal, collective and extremist violence. Commonw Secr Public Heal Wales
8. Kassebaum NJ, Arora M, Barber RM et al (2016) Global, regional, and national disability-adjusted life-years (DALYs) for 315 diseases and injuries and healthy life expectancy (HALE), 1990–2015: a systematic analysis for the global burden of disease study 2015. Lancet 388(10053):1603–1658. https://doi.org/10.1016/S0140-6736(16)31460-X
9. Wang H, Naghavi M, Allen C et al (2016) Global, regional, and national life expectancy, all-cause mortality, and cause-specific mortality for 249 causes of death, 1980–2015: a systematic analysis for the global burden of disease study 2015. Lancet 388(10053):1459–1544. https://doi.org/10.1016/S0140-6736(16)31012-1
10. WHO. Preventing youth violence: an overview of the evidence. http://www.who.int/violence_injury_prevention/violence/youth/youth_violence/en/. Published 2015

References

11. Boers K (2019) Delinquent behavior over the life course. Monatsschrift für Kriminologie und Strafrechtsreform 102(1):3–42. https://doi.org/10.1515/mks-2019-0004
12. Heinz W (2017) Kriminalität Und Kriminalitätskontrolle in Deutschland – Berichtsstand 2015 Im Überblick (Crime and Crime Control in Germany - Overview of the 2015 Reporting Status). http://www.ki.uni-konstanz.de/kis/
13. Frieze IH, Newhill CE, Fusco R (2020) Dynamics of family and intimate partner violence, Chap. 4. Springer Nature Switzerland AG, Cham
14. Devries KM, Mak JYT, Garcia-Moreno C et al (2013) The global prevalence of intimate partner violence against women. Science (80-)340(6140):1527–1528. https://doi.org/10.1126/science.1240937
15. Abrahams N, Devries K, Watts C et al (2014) Worldwide prevalence of non-partner sexual violence: a systematic review. Lancet 383(9929):1648–1654. https://doi.org/10.1016/S0140-6736(13)62243-6
16. Bundeskriminalamt (BKA). Partnerschaftsgewalt, Kriminalstatistische Auswertung—Berichtsjahr 2019 (German Federal Criminal Police Office: Partner Violence, Criminal Statistical Analysis—Reporting Year 2019). https://www.bka.de/DE/AktuelleInformationen/StatistikenLagebilder/Lagebilder/Partnerschaftsgewalt/partnerschaftsgewalt_node.html. Published 2019. Accessed 23 November 2020
17. UNICEF. A familiar face—violence in the lives of children and adolescents. https://data.unicef.org/resources/a-familiar-face/. Published 2017
18. Unicef (2017) Unicef-report "a familiar face. violence in the lives of children and adolescents." https://data.unicef.org/wp-content/uploads/2017/10/EVAC-Booklet-FINAL-10_31_17-high-res.pdf
19. UNICEF. Violence against children. https://www.unicef.org/protection/violence-against-children. Published 2020. Accessed 18 March 2021
20. Berthold O, Frericks B, John T, Clemens V, Fegert JM, von Moers A (2018) Abuse as a cause of childhood fractures. Dtsch Aerzteblatt Online. https://doi.org/10.3238/arztebl.2018.0769
21. Frieze IH, Newhill CE, Fusco R (2020) Child maltreatment: physical, emotional and sexual abuse of children and child neglect, Chap. 6. Springer Nature Switzerland AG, Cham
22. Fearon J, Hoeffler A (2014) Benefits and costs of the conflict and violence targets for the post-2015 development agenda.https://doi.org/10.1017/CBO9781107415324.004
23. Institute for Economics & Peace (2018) The economic value of peace 2018: measuring the global economic impact of violence and conflict. http://visionofhumanity.org/reports/
24. Stockholm International Peace Research Institute (SIPRI). Global military expenditure sees largest annual increase in a decade—says SIPRI—reaching $1917 billion in 2019. SIPRI.org. https://sipri.org/media/press-release/2020/global-military-expenditure-sees-largest-annual-increase-decade-says-sipri-reaching-1917-billion. Published 2020. Accessed 28 April 2020
25. UNHCR (2019) Global trends, forced displacement in 2018. Genf. https://www.unhcr.org/5d08d7ee7.pdf
26. Pinker S (2011) The better angels of our nature: why violence has declined, Chaps. 2, 3. Viking Books Adult
27. Rosner M, Ritchie H Homicides—how have homicide rates changed over time? Our world in data. https://ourworldindata.org/homicides. Published 2019
28. Elkins C (2020) The "moral effect" of legalized lawlessness, violence in britain's twentieth-century empire. In: Dwyer P, Micale MS (eds) On violence in history, vol 1. Berghahn Books, New York, pp 293–338
29. Micale MS (2020) What pinker leaves out. In: Dwyer P, Micale MS (eds) On violence in history, vol 1. Berghahn Books, New York, pp 466–508

Chapter 4
Why is the Tendency to Violence a Human Trait?

Violence is not the work of evil forces, although it is sometimes tempting to assume that it is. It is the product of a mentality shaped by genetic, biographical and current social influences. Our genetic make-up, which includes a predisposition to aggression and violence, is the result of many millions of years of phylogenetic development. This phylogenetic legacy will also have a decisive influence on the future course of our history.

Aggression and Violence as a Result of Human Evolution: Phylogenetic Causes

Human psychological traits, like physical traits, are the product of a long phylogenetic development. In several million years of the development of mankind not only physical, but also psychological characteristics asserted themselves that were advantageous in the struggle for existence for the continuation of a species. Darwin was the first to point this out in 1859 [1].

It should be emphasized here that social Darwinism, which is based on this insight, is not a scientific point of view, but a judgmental ideology that classifies different human groups or races as superior or inferior, up to racist superiority thinking and the idea of a master race. Our recent history has shown that this ideology had devastating consequences, especially at the time of National Socialism.

The environmental conditions that the precursors of *Homo sapiens* had to deal with were not only climate, food supply, predators and infectious agents, but also in particular other humans competing for food, territories and sexual partners. Here, those were at an advantage who, due to physical superiority, higher intelligence, better communication and social interaction and the resulting superior group organization, were able to assert themselves and thus pass on their genes to their offspring more easily than the losers. Decisive for survival in the early phases of human development and thus for reproductive success was not only the natural but also the social environment. Cohesion within a clan, village or tribe with successful aggression against other competing groups increased the probability of survival.

Elementary psychological traits—which include aggression and violence as well as compassionate and prosocial attitudes—can only unfold on the basis of an evolutionary, inherited personality disposition that makes situationally appropriate behavior possible. This does not diminish the importance of the influences of the individual life history and the psychosocial environment on personality development, which is especially formative in the first years of an individual's life, but can also shape personality in later phases of life. As the end product of phylogenesis, however, the basic genetic make-up of humans provides the framework within which social and cultural influences can shape our character traits. To illustrate this with the example of a musical instrument: the genes are, so to speak, the claviature on which the social environment can play. However, only music that is made possible by the arrangement of the given keys and the sound quality of the strings can be created. Outside the genetically set basic conditions, which vary from individual to individual, environmental influences remain ineffective.

The evolution theory of the emergence of elementary mental characteristics such as intelligence and aggressiveness, but also peacefulness and friendliness, implies that the disposition for this in *Homo sapiens* gradually developed over tens of thousands of generations through natural selection of the underlying genetic equipment. This general principle of phylogenesis, based on variation, selection and inheritance, was called natural selection by Darwin [1]. This principle applies to elementary mental as well as to physical characteristics.

Changing Character Traits by Selective Breeding

The natural selection of mental traits during phylogenesis over many generations proceeds according to similar principles to the formation of different character traits of animals by selective breeding. Therefore, the principles of evolution of mental traits and behavioral patterns can be illustrated by the example of selective breeding of animals with the desired characteristics. The difference from natural selection is that here it is not the natural or social environment that determines the chance to reproduce, but the selection of animals for reproduction by the breeder according to certain criteria in order to achieve a special expression of the desired characteristics in the offspring after several generations.

As early as the first half of the last century, experiments with laboratory rats demonstrated that their "intelligence" as measured by their learning behavior in a maze test could be strengthened or weakened by selective breeding over a few generations [2]. The offspring of the rats that performed best in the maze and the offspring of those that performed worst were each mated separately. After only eight generations, two strains of rats were present that differed significantly in their maze intelligence. To rule out the possibility that the offspring's good performance was learned from their parents, the young were alternately raised by adoptive parents from the other group. The offspring of the intelligent rats made significantly fewer mistakes than their own offspring even when adopted by less learning parents; the offspring of the "dumb" rats performed worse in the maze even when adopted by the "intelligent" parents. Thus, in this experiment, genes were selected for better (or worse) learning behavior.

Besides intelligence, many other personality traits are subject to genetic influences. For example, aggressive characteristics of fighting dogs can be strengthened or reduced again by selective breeding over generations [3]. By selective breeding of mice, after four or five generations a strain was already reached that was clearly more aggressive than an ordinary laboratory mouse [4].

Even peaceful behavior can result from selective breeding. In his experiments with Siberian foxes, which are otherwise aggressive and not to be accustomed to the proximity of human, the Russian biologist Belyaev [5], through selection of tame specimens over 35 generations, achieved that they endured the proximity of humans willingly and even showed themselves cuddly like domestic animals.

Parallel Development of Aggression and Sympathy During Evolution

In the several million years' development of mankind (humans and apes separated in the phylogenesis about seven million years ago, which corresponds to about 250,000 to 300,000 generations), our today's disposition to aggressive-violent parallel to peaceful fellow-human behavior developed by natural selection of the underlying genetic equipment. Both—aggressiveness and sympathy—are the essential building blocks of the mentality of *Homo sapiens*, which continue today to determine our social life in different forms alongside or against each other.

In a study seeking to find out how the mentality of entire groups of people affects their ability to survive in the course of evolution, various interaction possibilities within the own group and with foreign groups over thousands of generations under living conditions in the early period of mankind's development were simulated by computer programs using analytical methods from game theory. It was found that the groups that had the greater chances of survival were those that showed the highest warlike attitude towards foreign groups combined with the most pronounced cohesion within the own group [6]. This statistical simulation study of conditions in prehistoric times over many generations also makes clear that the genetic conditions for cohesion and feeling of solidarity within the own clan or the own tribe on the one hand, connected with aggressive behavior towards foreign groups on the other hand, anchored themselves as phylogenetic universals in our hereditary material. The genetic preconditions for successful group behavior brought about by evolutionary biology were passed on to the descendants and have remained a basic mentality of human coexistence anchored in us since time immemorial.

This inherited primal mentality, as a kind of collective unconscious, still determines the xenophobic behavior of human groups today, especially when compassion, empathy, feelings of foreign value, and appreciation of other ways of thinking and cultures are insufficiently conveyed through education and socialization.

Phylogenetic antecedents of human group aggression can be observed in chimpanzees, which shared common ancestors with the human species in evolutionary history. In her book "*In the Shadow of Man*" (1971) Goodall [7] describes the group behavior of these primates. She observed chimpanzee groups in their natural habitat in Africa over a period of years and even gradually placed herself in these groups until she was accepted as a clan member. She impressively described that within the chimpanzee groups not only did peaceful coexistence prevail after the establishment of the social hierarchy, but also hordes of adult male chimpanzees regularly roamed the boundaries of their territory. As soon as they encountered another monkey group, they attacked it until it was driven away, often resulting in lethal confrontations.

In an article entitled "The phylogenetic roots of human lethal violence" published in *Nature* by a Spanish research group [8], it was proven that in all investigated species of mammals (including humans), lethal aggression against own conspecifics occurs most frequently in primates and humans. As in primates, about two percent of the deaths in the course of the evolution of mankind can be explained by acts of killing own conspecifics. This death rate can, however, vary considerably depending on the respective cultural conditions and social framework constellations. It has decreased drastically in Europe since the late Middle Ages. In Europe, the incidence (cases per 100,000 persons per year) of homicides is currently about 1; that is, 0.001% of the population dies annually in Europe today by murder or homicide.

Why did Prehumans and Early Humans Disappear?

Looking back at the evolutionary history of mankind, the question arises why the numerous precursors of *Homo sapiens* disappeared in the course of phylogenesis. These include *Australopithecus* (which lived between 4 and 2 million years ago), *Homo habilis* (2–1.5 million years ago), *Homo erectus* (1.8 million–40,000 years ago), and Neanderthal man (200,000–30,000 years ago). *Homo sapiens* lived together with the latter for several thousand years after his immigration to Europe until Neanderthal man finally disappeared. The assumption is obvious here: these early humans perished not only because of climatic or other adverse living conditions, but also because they were wiped out by superior races of the genus *Homo* in the course of rival group fights and ethnic conflicts in the manner of pogroms over many generations. It is therefore likely that in *Homo sapiens* versus Neanderthal, and before that *Homo erectus* versus *Homo habilis*, one race of early forms of humans versus another, similar to what continues to be seen today both in primate group fights and in *Homo sapiens*, warlike clashes, pogroms and racial hatreds displaced and eliminated their competitors. Willingness to use violence against other people who did not belong to one's own group, combined with superior physical abilities, higher intelligence and the resulting more effective group organization, led over thousands of generations to the displacement of rivals for territory and food resources and to the spread of the genes of the victors, whose mentality was thus passed on.

There is extensive archaeological evidence of warlike conflicts in the early history of mankind. The American authors Keeley and Pinker reported on violence in prehistoric tribal societies [9, 10]. The results of archaeological excavations in Ukraine, France, Sweden, Asia, Africa and America testify to this. The oldest known site of such a prehistoric massacre, which took place 12,000 to 14,000 years ago, is on the Nile River in northern Sudan (Jebel-Sahaba). Skeletons of 40 men, women and children were found there, half of whom had stone bullets stuck in their bones. In recent years, other findings from early and prehistoric times have also yielded extensive evidence of prehistoric individual and collective acts of violence. An impressive compilation of such evidence was shown in the special exhibition "War—an Archaeological Search for Traces" at the State Museum of Prehistory in Halle, Germany (2015/2016) [11]. These included weapons from the Stone Age, such as the disc club, which was unsuitable as a hunting instrument but could be used to slay other people (Figs. 4.1 and 4.2), as well as battle scenes from the Neolithic period in which our prehistoric ancestors raced against each other with bows, arrows and lances (Fig. 18.1, page 175), and mass graves of slain people from the period 10,000–5000 BCE (Fig. 4.3) [12, 13].

Probably the best-known prehistoric figure who became a victim of violence more than 5000 years ago is "Ötzi", who was found as an ice mummy in a melting glacier in southern Tyrol. An arrowhead was stuck in his shoulder.

Fig. 4.1 Stone Age disc club. *Source* Meller and Schefzik, with permission of the Landesamt für Denkmalpflege und Archäologie Sachsen-Anhalt. see Ref. [11]

Fig. 4.2 Oldest documented murder in human history, 430,000 years ago, Spanish Sierra de Atapuerca [11]. Fractures in the area of the forehead are clearly visible. Cranium 17 from Sima de los Huesos, Javier Trubeba/Madrid Scientific Films

Fig. 4.3 Mass grave of slain people, near Halberstadt, Germany, ca. 5000 BCE. *Source* Meller and Schefzik, with permission of the Landesamt für Denkmalpflege und Archäologie Sachsen-Anhalt. see Ref. [11]

Decrease in Violence with Increasing Civilization?

There is some evidence suggesting that as civilization increased, the level of lethal violence decreased. The percentage of deaths in prehistoric archaeological sites due to interpersonal violence is reported by Pinker [14] and Keeley [9] in the range of 10–30%, and between 20 and 60% for the Native Americans before the arrival of the Europeans, as well as for the indigenous peoples in Amazonia, New Guinea, Australia and also for the Inuits. In contrast, deaths from wars in the European states and in the US in the twentieth century are in the range of 1–6%. The proportion of violent deaths of the world population in the twentieth century due to warlike conflicts, including the two world wars, has been estimated at 3% [10].

Among the indigenous peoples of Latin and Central America, the Incas, Aztecs and Mayans, signs of violent injury were found in 13% of the inhabitants living there as hunter-gatherers; among the city dwellers, 2.7% bore traces of violent injury. This, like the figures cited earlier, indicates that the risk of becoming a victim of a violent act decreases as civilization increases [15]. Keeley [9] and Pinker [16] conclude that even today in nongovernmental societies, deadly intertribal clashes are almost commonplace; for example, among the head-hunters of New Guinea and Borneo, the Maasai and Zulu warriors of Africa, and the inhabitants of the Amazon basin. The murder rate there is many times higher than in state-organized societies. The average for violence-related deaths in nongovernmental societies has been estimated at 500 per 100,000 population. Murder rates in the least violent non-state societies s were reported to be 100 per 100,000 population for the Inuit in the Canadian Arctic, 30 for the Kung in South Africa, and also 30 for the Semai in Malaysia [16]. In comparison, murder and manslaughter killed in recent years 6 per 100,000 people in the US and about 1 per 100,000 in Western Europe.

The depictions of Keeley and Pinker were criticized because they are said to have overestimated the violence of our ancestors on the basis of inaccurate source analyses and underestimated that of later generations [17–19]. This also gives the impression that mankind was exclusively afflicted by murder and manslaughter in our history and prehistory. The findings from all archaeological excavations taken together would yield a rate of violent deaths of less than 15%. It has been argued that modern societies and states are no less violent than their historical predecessors [20]; from antiquity to the present day, armies have grown ever larger, the technology of weapons and destruction ever more sophisticated, and thus the number of deaths in warlike conflicts ever increasing [21]; the first half of the twentieth century, it has been argued, was the deadliest in the history of the world [22]. It has also been pointed out that at the same time as the decrease of violence in Central and Western Europe, there was a massive increase of imperialist violence by Europeans in the colonies of the British Empire and those of other European countries, with many millions of victims among the indigenous population in Africa, Asia, America, Australia and New Zealand [23, 24].

The high presence of lethal violence throughout history should not obscure the fact that, in addition to the predisposition to violent behavior, peaceful coexistence was also subject to positive phylogenetic selection. This unfolded in different groups and societies, however, in different forms. The development of human history was characterized not only by warlike conflicts, but also by interpersonal cooperation as the most important prerequisite for the cohesion and continuity of families, clans, tribes and communities. There are also considerable differences between the various indigenous populations with regard to violent and peaceful ways of life.

The extent to which aggressive-warlike behavior on the one hand and peaceful coexistence on the other can develop in different directions during phylogenesis is illustrated by the example of chimpanzees and bonobos. Among the apes, both are our closest genetic relatives. In contrast to chimpanzees, bonobos are very peaceful animals. We are about equally related to both primate species and we have about 98% of DNA in common with them [25]. Both ape species live in central Africa, and the question is why the smaller, weaker and more peace-loving bonobos were not wiped out by the more aggressive chimpanzees in the course of evolution. The reason is simple: chimpanzees live north of the Congo River, bonobos south of it; the river is an impassable barrier for these apes.

Phylogenesis of Prosocial Behavior

The major counterparts of violence are prosocial traits that inhibit violent acts or prevent them from occurring at all. These include compassion, sympathy, benevolence, helpfulness, fairness, sense of justice—everything that is considered moral behavior. These behavioral categories are not usually considered from the point of view of the natural sciences, but rather fall into the realm of ethics, religion, philosophy and social sciences. Such human characteristics, however, did not develop out of nothing or by higher inspiration, but our disposition to mutual help and cohesion is just as much a product of a human development spanning hundreds of thousands of generations as our disposition to violence.

In his book 'A Natural History of Human Morality', Tomasello [26] has argued in detail that what we consider to be moral action has been favored by phylogenesis in the course of evolution because it was important for the survival of humankind. Prosocial traits such as trust, responsibility, obligation, commitment, guilt and blame, fairness, and compassion evolved in parallel with antisocial traits and the predisposition for aggression and violence. These two elementary, archaic components of our character continue to co-exist competitively in each individual, in the group behavior of humans, and in entire societies today.

In case of danger, prosocial mentalities for one's own group favored it in the fight against others in whom these characteristics were less pronounced. Altruistic commitment to one another, putting aside self-interest in favor of the group or, more generally, moral qualities were just as subjected to phylogenetic selection as aggressiveness and propensity to violence [26]. Cooperation requires adherence to common rules and the willingness to put aside self-interest in favor of group interests. A group which consists of selfish members without mutual helpfulness will assert itself in the existence fight less well than a cooperative mutually supporting community with better group organization and therefore has a smaller chance to pass copies of its genes predisposing to the antisocial mentality on to the offspring [6].

The two poles—'aggression and violence on the one hand' and 'altruism, empathy and prosocial behaviors' on the other—have developed differently in the multitude of past and present cultures and are quite unequally developed among some primitive peoples still living today. The 'Kung Bushmen' of southern Africa, who have a low birth rate and rely on group hunting and community cohesion for survival in adverse environmental conditions, have a markedly socially cooperative climate, with very low rates of violence among themselves, whereas the Munduruku Indians of the Amazon Basin have a violence and homicide rate of 30% of all male deaths; the higher a man's propensity for violence there, the higher his social status and the higher his chances of reproduction [27].

The assumption that intergroup conflict over many generations has contributed significantly to a genetic disposition not only to violence but also to altruistic behavior toward one's own group by no means rules out the possibility that behavioral patterns and mental attitudes have also been transmitted culturally from generation to generation. In addition to the transmission of genes predisposing to prosocial behaviors, the transmission of cultural values, which include religious attitudes, between generations is essential. However, the possibilities for the development of particular mentalities, shaped by phylogenesis, provide the framework within which cultural influences can take effect.

The evolutionary origin of altruism is also visible through observation of great apes: chimpanzees and other primates already show simple prosocial relationships with certain members of their group. According to Tomasello [26], the first step of an evolution to early forms of morality took place hundreds of thousands of years ago, when a change in ecology forced early humans to forage with partners or groups in order to survive. This required the definition of common goals, a common intentionality. The prerequisite for this was the following of common rules, and furthermore, compassion beyond relatives and friends also for other cooperating partners. The survival of the group was tied to a collective success, which required a coordinated group organization and cooperativity to achieve a common goal. This required mentalities that had moral awareness in addition to intelligence and communication skills [26].

There is also evidence in chimpanzees that they support each other and have compassion for those they help, which is associated with an increase in the mammalian bonding hormone oxytocin. This is true not only in the case of mutual grooming, but also in the case of such primates sharing food [28]. Thus, it is reasonable to assume that the last common ancestors of humans and apes already had prosocial traits, which parallels aggressive behavior to defend social status within the group and against outside groups in the struggle for territory and resources.

References

1. Darwin C (1859) On the origin of species by means of natural selection, or the preservation of favoured races in the struggle for life. John Murray, London
2. Tyron RC (1940) Genetic differences in maze-learning ability in rats. Yearb Natl Soc Stud Educ 39:111–119
3. Breed MD (2003) What is the basis for aggression in dogs? animalbehavioronline.com. http://www.animalbehavioronline.com/dogaggression.html. Published 2003. Accessed 31 Mar 2021
4. Cairns RB, MacCombie DJ, Hood KE (1983) A developmental-genetic analysis of aggressive behavior in mice: I. Behavioral outcomes. J Comp Psychol 97(1):69–89. http://www.ncbi.nlm.nih.gov/pubmed/6603330
5. Dugatkin LA (2018) The silver fox domestication experiment. evolution-outreach.biomedcentral.com. https://evolution-outreach.biomedcentral.com/articles/https://doi.org/10.1186/s12052-018-0090-x. Published 2018. Accessed 31 Mar 2021
6. Choi J, Bowels S (2007) The coevolution of parochial altruism and war. Science (80-) 318:636–640
7. Goodall j (1971) in the shadow of man. William Collins Sons & Co., London
8. Gómez JM, Verdú M, González-Megías A, Méndez M (2016) The phylogenetic roots of human lethal violence. Nature 538(7624):233–237. https://doi.org/10.1038/nature19758
9. Keeley L (1996) War before civilization: the myth of the peachful savage. Oxford University Press
10. Pinker S (2011) The better angels of our nature: why violence has declined, Chap. 2. Viking Books Adult
11. Meller H, Schefzik M. (2015) Krieg—Eine Archäologische Spurensuche. Begleitband Zur Sonderausstellung Im Landesmuseum Für Vorgeschichte Halle (War—an archaeological search for traces. Accompanying volume to the special exhibition in the State Museum of Prehistory Halle (Saale).). Landesamt für Denkmalpflege und Archäologie Sachsen-Anhalt, Halle(Saale). https://www.researchgate.net/publication/284142322_H_MellerM_Schefzik_Hrsg_Krieg_-_eine_archaologische_Spurensuche_Begleitband_zur_Sonderausstellung_im_Landesmuseum_fur_Vorgeschichte_Halle_Saale_6_November_2015_bis_22_Mai_2016_Halle_Saale_2015
12. Wild EM, Stadler P, Häußer A et al (2004) Neolithic massacres: Local skirmishes or general warfare in Europe? Radiocarbon 46(1):377–385. https://doi.org/10.1017/s0033822200039680
13. Meyer C, Lohr C, Gronenborn D, Alt KW (2015) The massacre mass grave of Schöneck-Kilianstädten reveals new insights into collective violence in Early Neolithic Central Europe. Proc Natl Acad Sci. https://www.pnas.org/content/112/36/11217. Published 2015. Accessed March 31 Mar 2021
14. Pinker S (2011) The better angels of our nature: why violence has declined, i. Viking Books Adult
15. Fry DP (2013) War, peace, and human nature: the convergence of evolutionary and cultural views. Oxford University Press
16. Pinker S (2011) The better angels of our nature: why violence has declined, Chap. 2. Viking Books Adult
17. Ferguson RB (2013) Pinker's list: exaggerating prehistoric war mortality. In: War, peace, and human nature. Oxford University Press, pp 112–131. https://doi.org/10.1093/acprof:oso/9780199858996.003.0007
18. Butler SM (2020) Getting medieval on Steven Pinker; violence and medieval England. In: Dwyer P, Micale MS (ed) On violence in history, vol 1. Berghahn Books, New York, pp 120–164
19. Dwyer P (2020) Whitewashing History; Pinker's (Mis)Representation of the enlightenment and violence. In: Dwyer P, Micale MS (ed) On violence in history, vol 1. Berghahn Books, New York, pp 206–249

References

20. Fibiger L (2020) The past as a foreign country: bioarchaeological perspectives on Pinker's "Prehistoric Anarchy." In: Dwyer P, Micale MS (ed) On violence in history, vol 1. Berghahn Books, New York, pp 40–76
21. Trundle M (2020) Were there better angels of a classical greek nature? Violence in classical Athens. In: Dwyer P, Micale MS (ed) On violence in history, vol 1. Berghahn Books, New York, pp 77–119
22. Roth R (2020) Does better angels of our nature hold up as history? In Dwyer P, Micale MS (ed) On violence in history, vol 1. Berghahn Books, New York, pp 339–376
23. Elkins C (2020) The "Moral Effect" of legalized lawlessness, violence in Britain's twentieth-century empire. In: Dwyer P, Micale MS (ed) On violence in history, vol 1. Berghahn Books, New York, pp 293–338
24. Micale MS (2020) What Pinker leaves out. In: Dwyer P, Micale MS (ed) On violence in history, vol 1. Berghahn Books, New York, pp 466–508
25. Prüfer K, Munch K, Hellmann I et al (2012) The bonobo genome compared with the chimpanzee and human genomes. Nature 486(7404):527–531. https://doi.org/10.1038/nature11128
26. Tomasello (2016) A natural history of human morality, Chap. 3. Harvard University Press
27. Raine A (2013) The anatomy of violence—the biological roots of crime, Chap. 1. Penguin Books, London
28. Tomasello M (2016) A natural history of human morality; Chap. 2. Harvard University Press

Chapter 5
Heritability of Aggressive Behavior

The influence of genetic make-up on our behavior has long been underestimated. However, it is quite substantial with regard to the tendency to violence as well as to other interpersonal traits. Which genes play a role and what does their activity depend on? What do genes do in the brain? Can genetic analyses predict the risk of violent acts?

Significance of Genes

Phylogenetic selection of elementary mental functions presupposes that at least a substantial proportion of the complex causal structure of a person's mental constitution is shaped by heredity and, like physical traits, is passed on from generation to generation. Unlike many physical traits, which are shaped exclusively by genetic make-up (such as blood type or eye color), mental traits are shaped both by genes as well as by early and late influences of the social environment. But to what extent can genes influence our behavior and feelings, and to what extent can our personality traits be shaped by our early and late social environment? In order to answer these questions, we will first briefly explain the functioning of a gene for a better understanding of the interaction between genetic make-up and environmental influences.

A gene consists of DNA base sequences that are the starting point for protein synthesis (gene expression) in the cell. Gene expression is the basis of forming the structure and function of all organs including the brain. The diversity of base sequences of DNA corresponds to the diversity of proteins synthesized from it, as well as to the diversity of cellular properties and organ functions. However, a gene does not prompt protein synthesis in a rigid, unalterable, lifelong manner, but is subject to regulatory influences of the environment.

The idea that a certain gene is responsible for a certain character trait is not accurate. Individual genes or the combination of different genes merely increase the probability for certain personality traits, such as antisocial behavior, aggressiveness and proneness to violence. Even most mental disorders will develop if certain constellations of the social environment inhibit or promote the activity of predisposing genes [1].

The Interplay of Genes and Environment: Epigenetics

At the molecular genetic level, the relatively new research field of epigenetics deals with the molecular fine-tuning of gene activity and environment. There are particular molecular mechanisms (in the nomenclature of molecular geneticists, these are DNA methylation, histone protein modification and microRNA regulation) that, depending on environmental influences, impede or facilitate the reading of the DNA code required for protein synthesis [1, 2] (Fig. 5.1). Several studies have demonstrated that psychosocial influences, such as early childhood maltreatment, stimulate such molecular epigenetic activities that impair DNA code reading for synthesis of proteins required in the brain for stress regulation and for nerve growth [1]. It is becoming more and more apparent that a variety of mental illnesses and personality disorders, including antisocial and aggressive behavior, are co-shaped not only by inherited genes but also by molecular epigenetic mechanisms that are the consequence of traumatizing early childhood experiences [3].

Therefore, the old controversial question of whether either the environment or genes are responsible for a person's mental reaction patterns, as well as for their pathological deviations, is superfluous. Genes, which synthesize proteins essential for the function of the brain and thus influence mental characteristics, are inhibited or strengthened in their activity by the epigenetically active molecules mentioned above, depending on environmental influences. Genes are therefore not inflexible

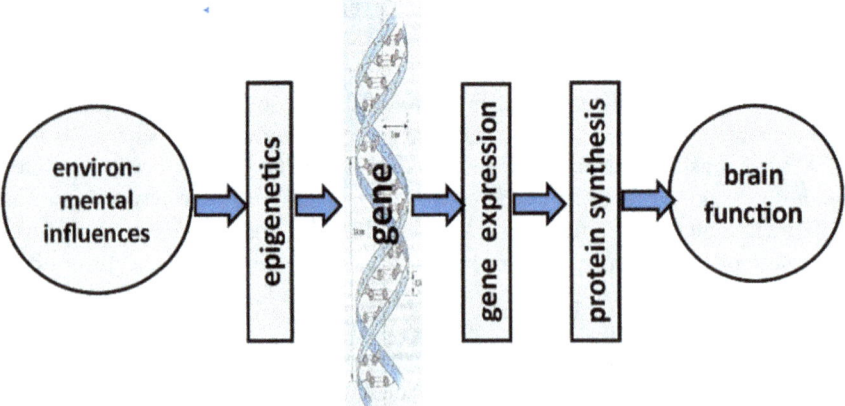

Fig. 5.1 Schematic representation of environmental influences on gene activity. Own illustration

actors that determine the course of our brain activities only on their own. They change their activity according to the respective environment; for mental equipment this is above all the psychosocial conditions present in childhood. It can thus be understood that genetic make-up can only explain a certain proportion of the causal variance of normal and pathological mental characteristics.

However, genetic disposition or environmental influences can have a different weighting in diverse mental characteristics or mental illnesses. For example, massive psychological trauma plays a decisive role in the set of conditions for post-traumatic stress disorders, while hereditary factors are predominant in bipolar disorders (formerly called manic-depressive disorders).

How Strong is the Influence of Genes? Twin and Family Studies

The interaction of genes and environment is well documented with the example of intelligence. Several studies show that about 50% of the determinants of intelligence are set by genes, the other half by environmental conditions [4]. The same is true for other psychological personality traits [5]. How pronounced the genetic variance is also depends on early social circumstances. For example, heritability for intelligent behavior is about 50–60% for IQ in children from families with high socioeconomic status. However, it is only about 20–30% for children with low socioeconomic status [6, 7]. A high social status thus allows the full unfolding of the genetically given possibilities of intelligence development, while a low status limits these influences. Likewise, the genes that dispose to aggressive and violent behavior reveal their effect only if a social environment suitable for this promotes these character traits.

In the 1980s, family, twin and adoption studies first demonstrated that both genetic factors and early family environment contribute to criminal behavior [8, 9]. Adopted male adolescents whose biological or adoptive fathers had been convicted of criminal offenses were studied in comparison with those whose biological and adoptive fathers had not. Adopted sons whose biological fathers had been delinquent showed a significantly higher rate of criminal behavior than sons of nondelinquent biological fathers, regardless of whether the adoptive father was a delinquent or not. The highest rate of convictions was found among those sons in whom both the biological and adoptive fathers had engaged in delinquent behavior. This finding has been confirmed several times; for example, by a nationwide adoption study in Sweden [10] that demonstrated that both violent and non-violent criminal behavior by the biological parents predicted a high percentage of similar behavior in their children, even when they grew up in an adoptive family.

The fact that a large number of character traits are hereditary has been demonstrated by comparing identical twins, whose genes are identical, with fraternal twins whose genetic equipment corresponds to that of normal siblings (Fig. 5.2).

Fig. 5.2 Genetics of delinquent behavior. The probability of aggressive-criminal behavior increases with the genetic degree of relatedness. For example, a concordance rate (degree of conformity) of 50% means that half of the respective population shows the same behavioral characteristics [8–10]. Own illustration

Such studies showed that antisocial and reckless behavior in particular, termed "psychopathy," is genetically determined for more than 50% of the causal spectrum.

In a summarizing statistical analysis (meta-analysis) of 103 studies, the heritability of aggressiveness was compared with rule-breaking non-aggressive behavior. Non-aggressive antisocial behavior was heritable to 48% of the causal variance, aggressive behavior to 65% [11].

Against the heritability of an inclination to violence it was objected that this is often accompanied by alcohol addiction, which itself has a strong genetic component and therefore it is the tendency to alcohol consumption rather than the resulting aggressiveness that is inherited [12]. The majority of the studies listed here speak against this, as does the evidence that aggressiveness already displayed in infancy and toddlerhood, in which alcohol consumption is not yet likely to play a role, has a considerable genetic causal share [13, 14].

Regarding the weighting of heredity and environment, twin and adoption studies have established that genes account for about half of the multiple sub-causes of antisocial behavior, with data varying between 30 and 70%, depending on the type of aggression and the population studied [15–17]. There appears to be a stronger heritability for overt aggression than for covert aggressive behavior. In children, a stronger genetic cause proportion (85%) was found for proactive aggression associated with lower empathy, compared with reactive aggression due to high emotional irritability (48%) [14, 18, 19].

Which Genes Play a Role?

The gene that was first suspected of predisposing to violent behavior is responsible for the metabolism of several transmitter substances in the nervous system (MAO-A gene) [20]. This gene can occur in variants with high or low activity. Low MAO-A gene activity is associated with a higher tendency to violence. However, this is only the case if the affected individuals had psychologically traumatizing experiences in childhood in the form of abuse. In individuals who had the MAO-A gene variant with high activity, aggressiveness was much rarer [21, 22]. It has been confirmed by several studies that a genetic disposition to aggressive behavior leads to later violent behavior only in interaction with early stressors in childhood; conversely, that in a genetic equipment without violence-disposing genes, comparable adverse life circumstances lead much less frequently to aggression [23]. Also, a meta-analysis summarizing data of 11,000 persons could prove a coincidence of low activity of the MAO-A genotype and aversive childhood experiences as conditional factors for later antisocial behavior [24]. The genotype with low MAO-A activity can lead to extremely violent behavior especially in combination with alcohol or drug consumption [25].

The importance of the MAO-A gene for aggressiveness has also been impressively demonstrated in animal experiments. Mice in which the MAO-O gene has been switched off are extremely aggressive and immediately attack their conspecifics [26].

The predisposition to certain behaviors is not determined by just one gene, but by the interaction of many genes. In a recently published review [26] summarizing all molecular genetic studies on aggressiveness in humans and laboratory animals, a ranking of 40 genes was established, many of which interact with other genes and contribute with different weights to aggressive behavior. At the top of this ranking list is the MAO-A gene.

In addition to the MAO-A gene, a gene responsible for serotonin transport in the brain (5-HTT gene) is also important. This gene has also been shown to predispose to violent behavior if the person was exposed to physical abuse as a child [27]. In addition, there are other genes that prompt to violent behavior both through interaction with each other and with the early psychosocial environment. Given the large number of genes that increase the likelihood of aggressiveness, a single gene is of relatively little importance. Moreover, the effect of individual genes is quite nonspecific. For example, the MAO-A gene increases not only the propensity for violence but also the risks for alcoholism; the 5-HTT gene increases the risk for violence and for attention deficit disorder [28].

What Do Genes Do in the Brain?

The influence of genes on the extremely complex structural and functional organization of the brain can only be presented here in a very simplified way and in only partial aspects. Genes that contribute to a person's greater or lesser propensity to violence are in particular those that regulate the metabolism of some of the brain's messenger substances (neurotransmitters). These include serotonin, dopamine and norepinephrine, which are important for the neurobiological mechanisms of aggressive behavior. The variant of the MAO-A gene associated with low expression of the enzyme monoamine oxidase (MAO-A), which degrades dopamine and norepinephrine in brain tissue, may cause overactivity of these drive-enhancing transmitters because of decreased degradation. Underactivity of the gene responsible for serotonin transport in the brain (5-HTT gene) is associated with deficient serotonin function, which facilitates aggressive behavior. Thus, these genes can promote violent behavior by an insufficient activity of serotonin or by an overactivity of dopamine and norepinephrine.

Can Gene Analyses Predict Dangerous Behavior?

In a meta-analysis [9] that evaluated the results of 185 studies, the question of whether dangerous behavior can be predicted on the basis of the presence of one of these gene variants was addressed. Based on the data, this was negated. The prediction of risk and extent of violent acts based on the analysis of single genes, for example, the MAO-A gene or the genes that are important for serotonin metabolism, is not (yet?) possible according to the current state of research. This is plausible because of the broad spectrum of psychosocial environmental influences and the multitude of genes that only in combination increase the risk for violence. Thus, genetic analyses are currently not useful for predicting the dangerousness of individuals with regard to therapeutic interventions or forensic issues.

Genes and Prosocial Traits

Not only antisocial but also prosocial character traits are influenced by genetic factors. Extensive twin and family studies have shown that prosocial behavior is significantly influenced not only by the social and family environment but also by genes [29]. The genetic share of the spectrum of causes is 50–60% for prosocial behavior in both sexes, in children and adults, and thus reaches approximately the same order of magnitude as the heritability for antisocial behavior and aggressiveness. For empathy, a genetic causal share of 40–50% was found. The prosocial genes also include a gene that is responsible in the brain for the receptor of the bonding hormone oxytocin (also called the cuddle hormone) [29].

Genes and the Future of Our Behavior

The human genetic equipment, which is the result of millions of years of phylogenesis of mankind, has not changed much in *Homo sapiens* during the last millennia. As since time immemorial, about half of the complex texture of causes of our mental equipment, be it antisocial or prosocial, or aggressive-violent or fellow-human traits, will remain firmly anchored in our genetic material for the next millennia. The genetically determined scope of our emotions and the manifold behavioral possibilities resulting from them will remain unchanged in their sum. There will not be a new type of human being in the near future. What can change quickly, on the other hand, are the psychosocial, social, and thus also political conditions that, via epigenetic mechanisms, slow down or stimulate the activity of our genes that predispose us to pro- or antisocial attitudes. This explains why groups of people up to entire communities of peoples, whose gene pool remains the same over centuries, can behave so differently at different times. Just compare Germany or Japan at the time of the Second World War with the same countries today.

Here we find the interface to the tangential points of genetics and sociology.

References

1. Binder EB (2019) Environment and epigenetics. Nervenarzt 90(2):107–113. https://doi.org/10.1007/s00115-018-0657-3
2. Ziegler C, Schiele MA, Domschke K (2018) Patho- and therapyepigenetics of mental disorders. Nervenarzt 89(11):1303–1314. https://doi.org/10.1007/s00115-018-0625-y
3. Gescher DM, Kahl KG, Hillemacher T, Frieling H, Kuhn J, Frodl T, Epigenetics in personality disorders: today's insights. Front Psychiatry. https://doi.org/10.3389/fpsyt.2018.00579
4. Plomin R, von Stumm S (2018) The new genetics of intelligence. Nat Rev Genet 19(3):148–159. https://doi.org/10.1038/nrg.2017.104
5. DiLalla D, Carey G, Gottesman I, Bouchard TJ (1996) Heritability of MMPI personality indicators of psychopathology in twins reared apart. J Abnorm Psychol 105:491–499
6. Tucker-Drob EM, Rhemtulla M, Harden KP, Turkheimer E, Fask D (2011) Emergence of a gene × socioeconomic status interaction on infant mental ability between 10 months and 2 years. Psychol Sci 22(1):125–133. https://doi.org/10.1177/0956797610392926
7. Harden KP, Turkheimer E, Loehlin JC (2007) Genotype by environment interaction in adolescents' cognitive aptitude. Behav Genet 37(2):273–283. https://doi.org/10.1007/s10519-006-9113-4
8. Mednick SA, Gabrielli WF, Hutchings B (1984) Genetic influences in criminal convictions: evidence from an adoption cohort. Science 224(4651):891–894. http://www.ncbi.nlm.nih.gov/pubmed/6719119
9. Vassos E, Collier DA, Fazel S (2014) Systematic meta-analyses and field synopsis of genetic association studies of violence and aggression. Mol Psychiatry. 19(4):471–477. https://doi.org/10.1038/mp.2013.31
10. Hjalmarsson R, Lindquist MJ (2013) The origins of intergenerational associations in crime: lessons from Swedish adoption data. Labour Econ 20:68–81. https://doi.org/10.1016/j.labeco.2012.11.001
11. Burt SA (2009) Are there meaningful etiological differences within antisocial behavior? Results of a meta-analysis. Clin Psychol Rev 29(2):163–178. https://doi.org/10.1016/j.cpr.2008.12.004

12. Bohman M (1978) Some genetic aspects of alcoholism and criminality. A population of adoptees. Arch Gen Psychiatry 35(3):269–276. https://doi.org/10.1001/archpsyc.1978.01770270019001
13. Helmsen J, Koglin U, Petermann F (2012) Emotion regulation and aggressive behavior in preschoolers: the mediating role of social information processing. Child Psychiatry Hum Dev 43(1):87–101. https://doi.org/10.1007/s10578-011-0252-3
14. Waltes R, Chiocchetti AG, Freitag CM (2016) The neurobiological basis of human aggression: a review on genetic and epigenetic mechanisms. Am J Med Genet Part B Neuropsychiatr Genet 171(5):650–675. https://doi.org/10.1002/ajmg.b.32388
15. Ferguson CJ (2010) Genetic contributions to antisocial personality and behavior: a meta-analytic review from an evolutionary perspective. J Soc Psychol 150(2):160–180. https://doi.org/10.1080/00224540903366503
16. Rhee SH, Waldman ID (2002) Genetic and environmental influences on antisocial behavior: a meta-analysis of twin and adoption studies. Psychol Bull 128(3):490–529. http://www.ncbi.nlm.nih.gov/pubmed/12002699
17. Cloninger CR, Sigvardsson S, Bohman M, von Knorring AL (1982) Predisposition to petty criminality in Swedish adoptees. II. Cross-fostering analysis of gene-environment interaction. Arch Gen Psychiatry 39(11):1242–1247. http://www.ncbi.nlm.nih.gov/pubmed/7138224
18. Tuvblad C, Raine A, Zheng M, Baker LA (2009) Genetic and environmental stability differs in reactive and proactive aggression. Aggress Behav 35(6):437–452. https://doi.org/10.1002/ab.20319
19. Bezdjian S, Raine A, Baker LA, Lynam DR (2011) Psychopathic personality in children: genetic and environmental contributions. Psychol Med 41(03):589–600. https://doi.org/10.1017/S0033291710000966
20. Brunner H, Nelen M, Breakefield X, Ropers H, van Oost B (1993) Abnormal behavior associated with a point mutation in the structural gene for monoamine oxidase A. Science (80-) 262(5133):578–580. https://doi.org/10.1126/science.8211186
21. Caspi A (2002) Role of genotype in the cycle of violence in maltreated children. Science (80-) 297(5582):851–854. https://doi.org/10.1126/science.1072290
22. Kim-Cohen J, Caspi A, Taylor A et al (2006) MAOA, maltreatment and gene–environment interaction predicting children's mental health: new evidence and a meta-analysis. Mol Psychiatry 11(10):903–913. https://doi.org/10.1038/sj.mp.4001851
23. Cicchetti D, Rogosch FA, Thibodeau EL (2012) The effects of child maltreatment on early signs of antisocial behavior: genetic moderation by tryptophan hydroxylase, serotonin transporter, and monoamine oxidase A genes. Dev Psychopathol 24(03):907–928. https://doi.org/10.1017/S0954579412000442
24. Byrd AL, Manuck SB (2014) MAOA, Childhood maltreatment, and antisocial behavior: meta-analysis of a gene-environment interaction. Biol Psychiatry 75(1):9–17. https://doi.org/10.1016/j.biopsych.2013.05.004
25. Tiihonen J, Rautiainen M-R, Ollila HM et al (2015) Genetic background of extreme violent behavior. Mol Psychiatry 20(6):786–792. https://doi.org/10.1038/mp.2014.130
26. Zhang-James Y, Fernàndez-Castillo N, Hess JL et al (2018) An integrated analysis of genes and functional pathways for aggression in human and rodent models. Mol Psychiatry 11:1655–1667. https://doi.org/10.1038/s41380-018-0068-7
27. Reif A, Rösler M, Freitag CM et al (2007) Nature and nurture predispose to violent behavior: serotonergic genes and adverse childhood environment. Neuropsychopharmacology 32(11):2375–2383. https://doi.org/10.1038/sj.npp.1301359
28. Retz W, Freitag CM, Retz-Junginger P et al (2008) A functional serotonin transporter promoter gene polymorphism increases ADHD symptoms in delinquents: interaction with adverse childhood environment. Psychiatry Res 158(2):123–131. https://doi.org/10.1016/j.psychres.2007.05.004
29. Ebstein RP, Israel S, Chew SH, Zhong S, Knafo A (2010) Genetics of human social behavior. Neuron 65(6):831–844. https://doi.org/10.1016/j.neuron.2010.02.020

Chapter 6
Neurobiology of Violence

Violence originates in the brain; that is, it is tied to the activity of certain centers in our brain, which in turn are subject to stimulating or inhibiting influences of surrounding brain circuits that depend on information from the social environment. The brain anatomical and brain-physiological preconditions for the development of violence are explained in this chapter.

Evidence of Aggression Centers in the Brain

In 1949, the Nobel Prize in Medicine was awarded to the Swiss physiologist Walter Hess for a discovery that profoundly changed our understanding of how the brain works and thus of the foundations of our state of mind. Hess used the cat as an experimental animal to demonstrate that electrical stimulation by means of fine electrodes inserted through the skull into nerve cell groups located deep inside the brain could trigger angry aggressive behavior at the click of a button, without any external reason for doing so (Fig. 6.1). Not only aggression but, depending on which cell group in this deep brain area was stimulated, expressions of the entire spectrum of archaic instincts, drives and emotions could be activated by weak electrical stimulation. In addition to aggression and anger, these included fear, flight, sexual behavior, sleep and a range of physical-vegetative responses regulated by these deep brain structures [1–4].

The arrangement and functioning of the anatomical structures of the brain region stimulated by Hess, the brain stem or more precisely the hypothalamus (for location in the brain, see Figs. 6.2, 6.3 and 6.4) are largely identical in all vertebrates up to and including humans. They have hardly changed during evolution. Therefore, it is not surprising that the same effects occurred after stimulation of deep brain structures in all mammalian species in which such experiments have been performed, including monkeys and humans [5–8].

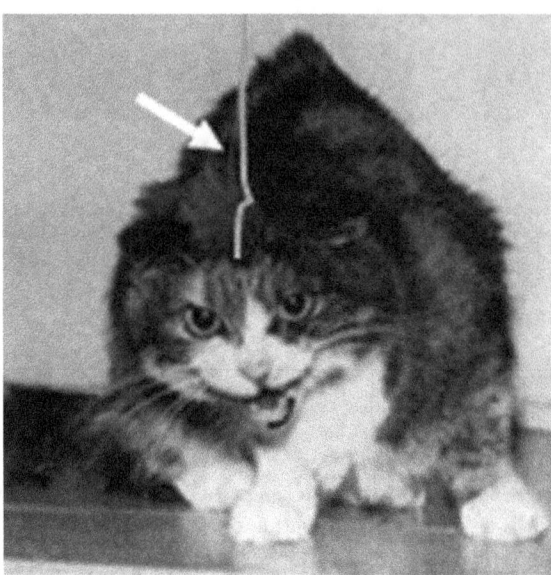

Fig. 6.1 Brain stimulation experiment by Hess on the cat (1943). Aggressive behavior can be induced by electrical stimulation of groups of nerve cells located deep within the brain. Arrow shows the wire of the electrode stimulating brain stem structures. *Source* Brown et al. [3]. Copyright Exp. Brain Res, Springer

In 1970, the American neurosurgeons Mark and Ervin reported on patients with raving states during epileptic seizures in whom all previous therapies had failed [5]. As a last choice, brain surgery was performed to eliminate the aggression-triggering cell groups in the brain. In these patients, electrical activity was evaluated via electrodes inserted into the brain during preoperative diagnostics in search of the seizure-triggering epileptic focus. When the amygdala (an important aggression-regulating center in the temporal brain) was activated by weak electrical stimulation via these electrodes, a massive outburst of aggression occurred in which the patients attacked everything that moved around them. When the electrical stimulation was stopped, this behavior ceased. By targeted destruction of these nerve cell groups in the amygdala, whose abnormal excitation was responsible for the violent outbursts, the epilepsy-related aggression attacks could be sustainably improved [5].

The results of classical electrical brain stimulation experiments have recently also been confirmed in animal experiments by newer optogenetic techniques by which minute groups of nerve cells can be selectively stimulated via light in the brain [9, 10]. By activating certain groups of nerve cells in this phylogenetically very old part of the brain stem, a wide range of instinctual actions, which in humans have a strong emotional component, can be elicited by direct stimulation.

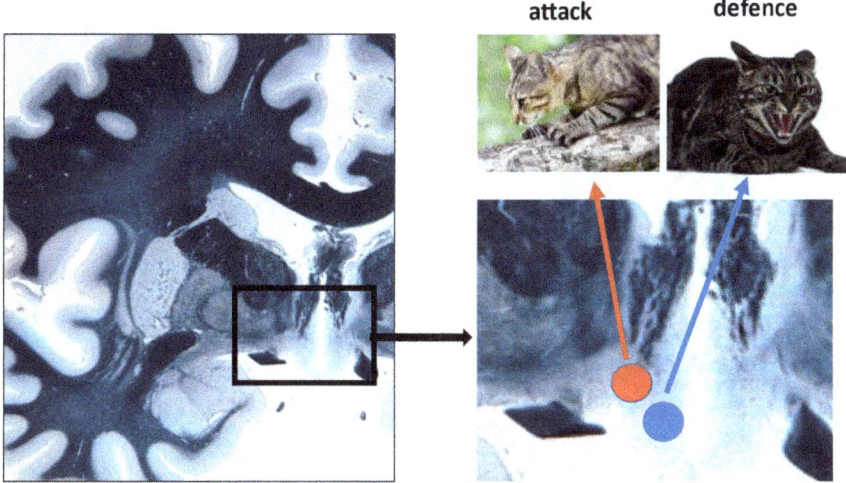

Fig. 6.2 Location of the reactive and proactive centers of aggression in the lower diencephalon. Left image: sectional plane through a human brain, frontal view; brain areas containing nerve tracts are colored black, areas containing nerve cells (gray matter) are colored light blue. Right image, lower half: detail from the left image showing the region of the brainstem (hypothalamus) with location of the aggression centers. Located close to the midline (blue) = nerve cell groups whose stimulation produces angry-aggressive defensive behavior, laterally (red) = nerve cell groups for proactive, planned aggression. Adapted from Elbert, T., Moran, J. K. & Schauer, 2017; see Ref. [11]

In addition, it could be shown that slightly laterally from the cell group in the diencephalon, by whose stimulation angry-aggressive behavior, as in defensive situations, could be triggered, another aggression center is located (Fig. 6.2). Its activation evokes planned and ordered aggressive behavior, as seen when a cat sneaks up on a prey animal. Thus, in this deep brain area there is both a center for angry, reactive aggression and another adjacent center for planned, proactive, "appetitive" aggression, which has a rewarding character when successfully performed.

The different forms of aggression thus have different neuronal correlates, which are located in close proximity to each other in a deep brain area (hypothalamus) (Figs. 6.2 and 6.4). Depending on which cell group is activated, violence can on the one hand be reactive, turning against an emerging danger or threat, and on the other hand it can also have a rewarding character. The latter is true for "appetitive" aggression and violence ranging from that directed at prey or exercised just for its own sake to its hedonistic and even sadistic manifestation [12].

These brain centers are not to be understood as units working in isolation from the rest of the brain, but they are central nodes of extensive neural circuits that are connected to other regions of deeper brain areas, to the limbic (emotion-relevant) system and related areas of the cerebral cortex [13]. The rage system is intensely connected to deep areas of the brainstem, through which it triggers the vegetative

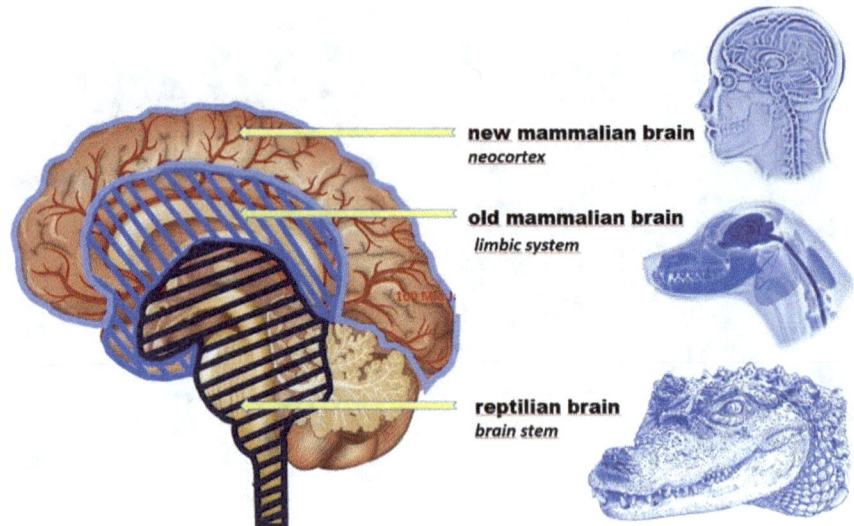

Fig. 6.3 Phylogenetic tripartite division of the human brain, according to MacLean (1952). The oldest part is the brain stem (phylogenetic age approx. 300 million years), which shows considerable similarities in structure and function between humans and lower vertebrates (e.g., reptiles). In the latter, however, the brain consists almost exclusively of this ancient part. The parts of the brain above it, the ancient mammalian brain (limbic system, phylogenetic age approx. 100 million years), and the neocortex (approx. 10 million years) appear later in the course of evolution as the brain becomes more complex and efficient. Own illustration. Partial images from Deposit photos No: 65265611/160557198/86104948

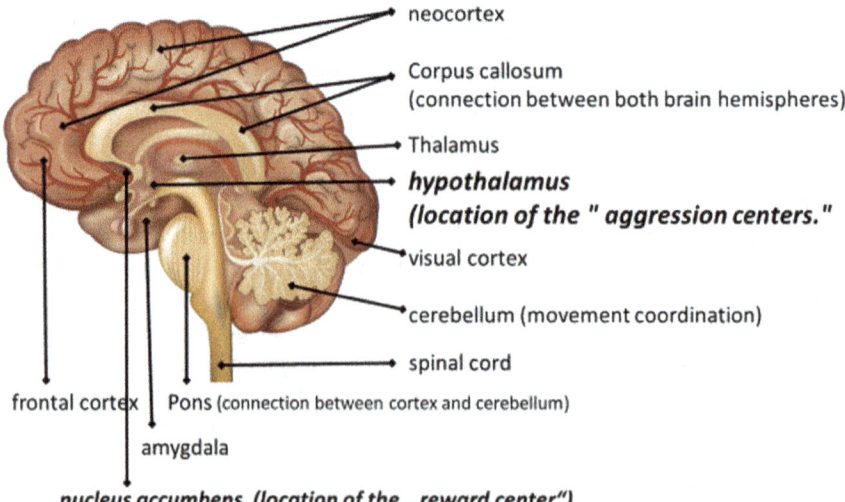

Fig. 6.4 Anatomy of the brain, midline view: location of aggression and reward centers. Own illustration. Partial images from Deposit photos No.: 86104948

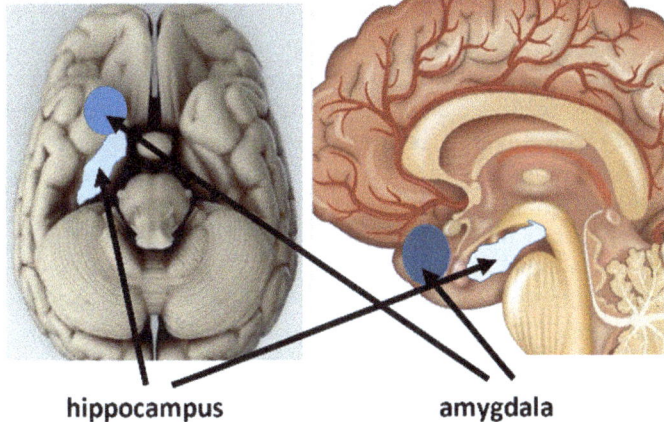

hippocampus   **amygdala**

Fig. 6.5 Middle and inferior brain views. Location of amygdala (dark blue) and hippocampus (light blue) inside the middle temporal brain. Own illustration. Partial images from Deposit photos No.: 86104948/1256818

responses associated with angry aggression, such as increases in heart rate, blood pressure elevation, sweating, pulse elevation and respiratory acceleration.

It was an obvious next step to inactivate these violence centers of the brain by means of neurosurgery in the case of the most serious violent offenders who proved to be resistant to all treatment approaches and who did not respond to any sociotherapeutic or psychopharmacological measures. However, the first such psychosurgical interventions performed in the 1940s and 1950s fell into disrepute because they irreversibly destroyed larger areas of brain tissue with significant side effects. With newer stereotactic methods, the small brain areas in the hypothalamus and amygdala can be targeted with millimeter precision and inactivated by small tissue damage or continuous electrical stimulation. Such stereotactic inactivation of the aggression centers in the diencephalon and amygdala has been performed mainly in the US and Japan on therapy-resistant violent offenders, with predominantly positive results [14–16].

Regulation and Control of the Aggression Centers in the Brain

If external circumstances influence our state of mind, it is only through the fact that they affect the interaction of neuronal networks in our brain via the sensory organs. The brain is often described as the most complex structure that exists in the universe. However, the principles of its structure and function are just as much the result of evolution over many millions of years as is the case for the construction principles and functioning of the other organs of the body. The complicated and diverse fine-tissue architecture of the brain and the complex interaction of billions of brain cells via the nerve tracts and switching points between the nerve cells (synapses) may appear extremely complicated at first glance. However, the basic principles of brain functioning are relatively easy to understand against the background of the tasks that the brain had to perform during the very long phylogenetic development up to *Homo sapiens*.

For a better understanding of the neuronal control of elementary emotional states, including aggression and violence, by the brain circuits responsible for this, the structure and the different functional levels of the central nervous system, as well as some brain anatomical details will be explained in the following section.

Phylogenetic Tripartition of Brain Structure and Function: Concept of the Limbic System

The brain can be divided into three major structural and functional parts that have different phylogenetic ages. The phylogenetically oldest part is the brain stem at the base of the brain. This part of the brain is already present in vertebrates that appeared very early in phylogeny, such as in fish and reptiles. It has therefore been called the reptilian brain (Fig. 6.3) [17]. This part of the brain contains nerve cell groups responsible for simple instincts essential for survival, such as food intake, sexual behavior, attack and flight, control of the autonomic nervous system, regulation of respiration, gastrointestinal tract, genital organs and circulation. Using this ancient part of the brain, simple vertebrates can respond in a reflex-like manner to key environmental stimuli, without greater behavioral flexibility. The structural elements and functioning of this archaic brain part, which appeared in evolution about 300 million years ago and remained indispensable because of its importance in regulating instincts and bodily functions essential for survival, have hardly changed in the course of phylogeny up to the human brain. Therefore, all instincts of the reptile brain are still present in us, even if unconsciously and (mostly) controlled by superior newer brain parts [18].

In the course of evolution in the ascending animal sequence, the brain stem was surrounded by a newer superordinated ring-shaped brain part, which is responsible for the control and regulation of the "reptile brain" by activation or inhibition. This part of the brain, which is added in early mammals in the history of the earth, is called the limbic system (=old mammalian brain, Fig. 6.3). Important central relay stations of the limbic system, the amygdala and the hippocampus, are located in the middle inferior temporal brain (Fig. 6.5). The amygdala plays a central role in the emotional valuation of what is perceived and in the generation of aggression and fear; the hippocampus is important for memory formation; everything that is emotionally relevant is deposited in the extensive memory areas of the neocortex for later retrieval through mediation by the hippocampus. Limbic structures control the activity of the brainstem neuronal centers for basic instincts, including the aggression cell groups in the hypothalamus, through several neural pathways by inhibitory or activating impulses. Thus, the limbic system is a kind of first guidance and control instance of the reptilian brain. Therefore, its reflexive primitive reactions, which are still present in simple vertebrates, can be better adapted to the present environmental conditions in the more highly developed animals.

With progressive phylogenesis, the more intelligent mammals, especially primates and above all humans, gained an extensive enlargement of the cerebral cortex, the neocortex (Figs. 6.3 and 6.4), which makes up the largest part of our brain. With the neocortex, which in humans is by far the largest of all living beings, we have a highly differentiated brain organ with a huge storage capacity for learned material and experience. In whales, elephants and dolphins, the absolute brain weight is even greater, but in relation to body size and body weight it is significantly smaller. In the neocortex, previous life experiences can be compared with the events of the current environment, integrated and associated. The most important task of the neocortex is the analysis of environmental events as well as action planning based on the stored previous experience.

Stages of Information Flow Through the Brain

The neocortex, after integration and association with past experience, passes the information it receives from the outside world to the limbic structures in the medial temporal lobe, the central stations of which are the amygdala and hippocampus.

The amygdala is responsible in particular for registering harmful environmental events that are detrimental to the individual. When it receives threatening information from the neocortex, it generates fear and/or aggressive behavior by activating, via specific nerve fiber pathways, the responsible nerve cell groups of the "reptile brain," or more precisely the fear or aggression cell groups of the hypothalamus located in the brainstem.

The enormous extent of the human cerebral cortex, which overlays the limbic system and the brainstem, as well as the principles of the brain anatomical stages of processing information from the sensory cortex (visual cortex, auditory cortex) via the association cortex and the limbic system to the hypothalamus can be seen in Fig. 6.6. A schematic representation of the sequence through which sensory impulses from the social environment are processed via the brain functional areas of phylogenetically different ages to the brain regions of archaic instincts is given in Fig. 6.7.

Under normal physiological conditions, the violence centers located in deep brain structures, which Hess [1, 2] was able to activate by direct electrical stimulation (see Fig. 6.1), are activated or inhibited by phylogenetically more recent superordinate parts of the limbic system and the neocortex. The most important cortical areas with violence-inhibiting brain functions are located in the frontal brain above the orbit and in the midline areas of the frontal lobe (cingulate cortex), as well as in the cortical

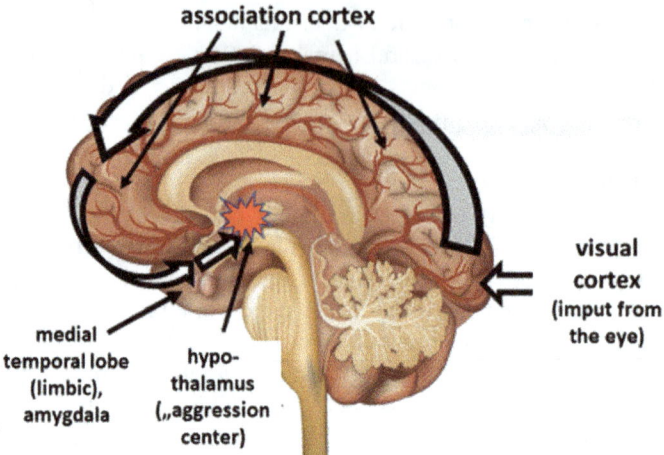

Fig. 6.6 View of the middle surface of the brain: simplified schematic representation of stages of information processing using the example of visual perception. Sensory impulses arriving in the visual cortex are transmitted to the association cortex and from there to limbic structures and amygdala. Depending on the information from the association cortex, the amygdala inhibits or activates the neurons of the "aggression center" in the hypothalamus Own illustration. Partial images from Deposit photos No.: 86104948

areas of the middle temporal brain surrounding the amygdala (see Figs. 6.6 and 6.8). It is therefore understandable that when these regions of the cerebral cortex are damaged, for example, by injuries, tumors or nerve cell loss in brain degenerative diseases, the threshold for aggressive and violent behavior is significantly lowered. This often results in such individuals reacting angrily and violently even to minor events [19–21].

Connection Between Violence and Reward Centers

The limbic system in a broader sense also includes the reward system, whose central station is the nucleus accumbens. This is located at the base of the brain in front of the hypothalamus (Figs. 6.4 and 6.8; see also Fig. 13.2, p. 108). Everything that is perceived as pleasant and is important for the survival of the species, such as food intake, sexual behavior and sense of belonging to one's own group, activates the reward center. Everything activating this center is associated with pleasure and well-being and is sought again; this also includes planned, successful "appetitive" aggression and violence.

The area for appetitive aggression is closely connected via nerve fibers to the central circuitry of the reward system, which is located in the immediate vicinity

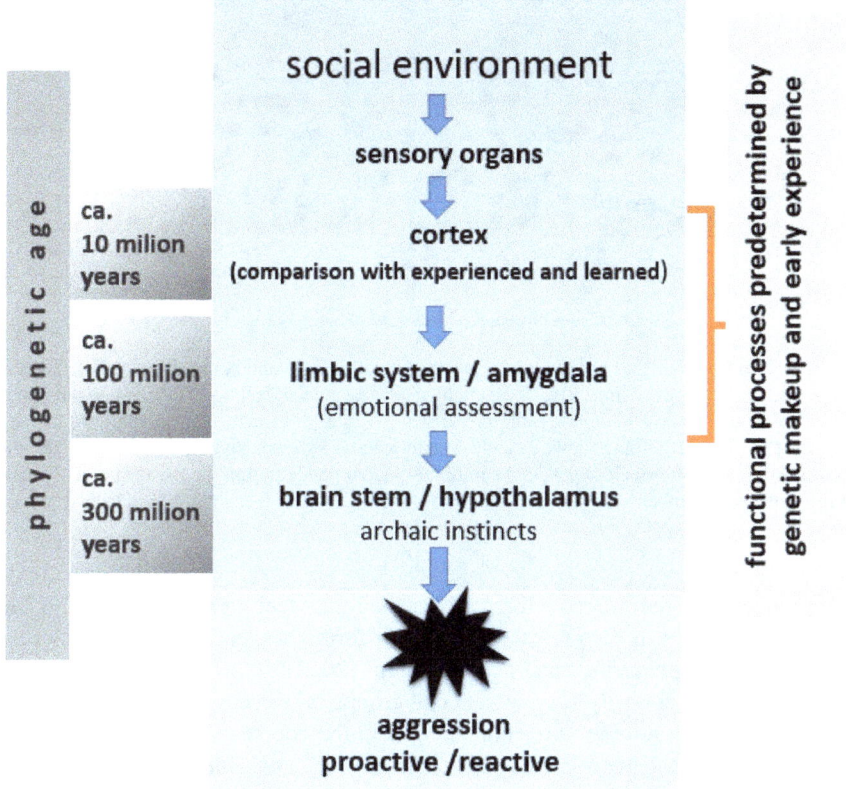

Fig. 6.7 Schematic representation of the different stages of information processing in the brain on the way to activating elementary emotions. Sequence of the information flow from the environment via the phylogenetically differently old brain functional areas to the activation or inhibition of the aggression centers. Own illustration

(Fig. 13.2). In psychopaths, for whom the use of violence is rewarding, the brain's reward center (nucleus accumbens) is overactive in certain experimental settings [22]; the same is true for *Schadenfreude* [23].

Neurobiology of Prosocial Behavior

By stimulating cell groups located in deep brain areas in the hypothalamus by means of special (optogenetic) light technology, depending on the activated cell group, not only can anger and aggression, fear and flight be triggered, but also affection (in animal experiments, grooming) [24, 25] or sexual behavior without an external occasion. Of particular importance for this is a group of cells located in the hypothalamus

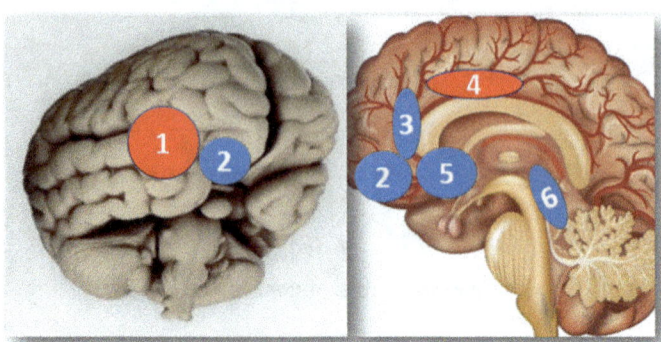

Fig. 6.8 Brain regions that become active during empathy (red) and compassion (blue). Source: adapted from Singer and Klimecki [30]. Empathy is witnessing the feelings of others; compassion is emotional engagement with others. Left: lateral view: 1. anterior insular region, 2. anterior inferior frontal brain. Right: midline view: 3. and. 4. anterior and superior cingulate cortex. 5. nucleus accumbens, 6. dopamine cell groups in the midbrain (dopamine is an important transmitter released in the nucleus accumbens during reward). Own illustration. Partial images from Deposit photos No.: 12568185/86104948

that produces the hormone oxytocin, also called the "feel-good," loyalty or cuddle hormone. Oxytocin is also released during all forms of pleasant human closeness and care; it strengthens the bond in partnerships [26, 27].

These phylogenetically very ancient cell groups in our reptile brain that are relevant for prosocial attitudes have not changed in the course of the evolution of our brain, just like the aggression-relevant cell areas. They are located right next to them in the hypothalamus. The neuronal correlates of the two great intrapsychic antagonists, aggression on the one hand, care and mutual affection on the other hand, are thus located in close proximity deep inside the brain in a very small region, in animals as well as in humans.

The control and regulation of the activity of the cell groups triggering positive prosocial feelings in the diencephalon occurs in a similar way to that of the aggression centers. The next higher-level instances are the cell groups of the thalamus [28] located above the hypothalamus (Fig. 6.4) and again the amygdala, whose middle part close to the cortex is connected via nerve fibers with the hypothalamic centers for caring and sexual behavior. For its part, the amygdala receives information from the limbic cortex (old mammalian brain in Fig. 6.3). The latter in turn is informed by the neocortex about what is going on in the environment [18] (Fig. 6.6).

Thus, very similar schemes of information flow in the brain apply to the control of sexuality, caring and interpersonal affection, and to the control of aggression and violence (Figs. 6.6 and 6.7).

Neurobiological Correlates of Ethics and Morals

Ethics and morality belong to the central topics of philosophy and religion, thus to areas that are classified as humanities. Ethical and moral values and actions based on them, however, did not emerge by themselves, but the preconditions for them are a result—just as for aggressive behavior—of a long phylogenetic development of mankind (see Chap. 4). Education and socialization can fill these preconditions of the brain created by evolution with concrete content—or may not.

Moral feeling, thinking and acting as the decisive basis of all prosocial behavior is, like all emotional and cognitive processes and ways of acting, bound to the function and undisturbed interaction of particular brain structures, which in turn are the result of phylogenetic selection. It is reasonable to assume that these are primarily the brain areas that control aggressive behavior, that is, the parts of the cerebral cortex and limbic system that inhibit violence-related cell groups in the brainstem.

However, prosocial behavior not only results from the suppression of aggressive impulses, but is above all the result of active interpersonal engagement, of affection, caring, fairness and altruism, thus of character traits that presuppose empathy and compassion. Special brain systems are also responsible for these human traits.

Brain Activity in Empathy and Compassion

The most informative and most frequently used method of examining the function (i.e., activation or inhibition) of brain regions during emotional or cognitive processes is functional magnetic resonance imaging (fMRI). If parts of the brain have to cope with certain tasks, one can measure the activation of the respective brain regions with millimeter precision using this equipment-intensive technique. Several studies of this kind have identified brain regions that become active during empathy, for example, when witnessing the pain of another person. These are located in the anterior insular region and the anterior cingulate cortex (anterior middle areas of the cortex) (Fig. 6.8) [29–32].

The term empathy has different meanings and is often used synonymously with interpersonal engagement [33]. Empathy in the narrow sense is limited to the ability to recognize and judge the feelings of others without being affected oneself. In addition, active sympathy, helpfulness, compassion and commitment to others [30, 34] can be seen as the most important characteristics of prosocial and moral behavior. If such a sentiment is evoked in subjects in fMRI experiments, brain regions in the anterior and inferior frontal brain become active, in contrast to empathy in the narrower sense; moreover, the brain's reward system belongs to those areas activated by compassion (Fig. 6.8) [29, 30]. However, like the aggression centers, these brain parts do not work as isolated units, but are central intersections of extensive neuronal networks that are connected to several other regions of the brainstem, limbic system and cortex [33].

The activation of the reward system by helpfulness and personal commitment towards others is the neurobiological reason why prosocial behavior is perceived as gratifying and pleasurable.

Individuals who are endowed with an inadequate capacity for compassion for others can improve this with appropriate mental training methods. This not only enables them to feel more positive interpersonal feelings. They also show better activation of the brain areas responsible for this in the middle, lower frontal brain, the limbic system and the reward system in fMRI studies [35]. The activation of the reward system is associated with a feeling of inner satisfaction and joy.

References

1. Hess W, Bruegger M (1943) Das subkortikale Zentrum der affektiven Abwehrreaktion. Helv Physiol Acta 1:33–52
2. Hess W (1949) Das Zwischenhirn. Syndrome, Lokalisationen, Funktionen. Benno Schwabe & Co., Basel
3. Brown JL, Hunsperger RW, Rosvold HE (1969) Defence, attack, and flight elicited by electrical stimulation of the hypothalamus of the cat. Exp Brain Res 8(2):113–129. https://doi.org/10.1007/bf00234534
4. Hunsperger RW, Bucher VM (1967) Affective behaviour produced by electrical stimulation in the forebrain and brain stem of the cat. In: Adey RW, Tokizane T (eds) Structure and function of the limbic system, vol 27, pp 103–127. Amsterdam/London/New York: Elsevier Publishing Company
5. Mark V, Erwin F (1970) Violence and the brain. Harper & Row, New York
6. Lin D, Boyle MP, Dollar P et al (2011) Functional identification of an aggression locus in the mouse hypothalamus. Nature 470(7333):221–226. https://doi.org/10.1038/nature09736
7. Delgado JMR (1964) Free behavior and brain stimulation. Int Rev Neurobiol 6:349–449. https://doi.org/10.1016/s0074-7742(08)60773-4
8. Haller J (2013) The neurobiology of abnormal manifestations of aggression—a review of hypothalamic mechanisms in cats, rodents, and humans. Brain Res Bull 93:97–109. https://doi.org/10.1016/j.brainresbull.2012.10.003
9. Steinberg EE, Christoffel DJ, Deisseroth K, Malenka RC (2015) Illuminating circuitry relevant to psychiatric disorders with optogenetics. Curr Opin Neurobiol 30:9–16. https://doi.org/10.1016/j.conb.2014.08.004
10. Falkner AL, Grosenick L, Davidson TJ, Deisseroth K, Lin D (2016) Hypothalamic control of male aggression-seeking behavior. Nat Neurosci 19(4):596–604. https://doi.org/10.1038/nn.4264
11. Elbert T, Moran JK, Schauer M (2017) Lust for violence: appetitive aggression as a fundamental part of human nature. e-Neuroforum 23(2). https://doi.org/10.1515/nf-2016-A056
12. Elbert T, Weierstall R, Schauer M (2010) Fascination violence: on mind and brain of man hunters. Eur Arch Psychiatry Clin Neurosci 260(S2):100–105. https://doi.org/10.1007/s00406-010-0144-8
13. Bartholow BD (2018) The aggressive brain: insights from neuroscience. Curr Opin Psychol 19:60–64. https://doi.org/10.1016/j.copsyc.2017.04.002
14. Bejjani BP, Houeto JL, Hariz M et al (2002) Aggressive behavior induced by intraoperative stimulation in the triangle of Sano. Neurology 59(9):1425–1427. https://doi.org/10.1212/01.WNL.0000031428.31861.23
15. Franzini A, Broggi G, Cordella R, Dones I, Messina G (2013) Deep-brain stimulation for aggressive and disruptive behavior. World Neurosurg 80(3–4):S29, e11-S29.e14. https://doi.org/10.1016/j.wneu.2012.06.038

16. García-Muñoz L, Picazo-Picazo O, Carrillo-Ruíz JD et al (2019) Effect of unilateral amygdalotomy and hypothalamotomy in patients with refractory aggressiveness. Gac Mex 155(91). https://doi.org/10.24875/GMM.M19000290
17. MacLean PD (1952) Some psychiatric implications of physiological studies on frontotemporal portion of limbic system (visceral brain). Electroencephalogr Clin Neurophysiol 4(4):407–418. http://www.ncbi.nlm.nih.gov/pubmed/12998590
18. Alcaro A, Carta S, Panksepp J (2017) The affective core of the self: a neuro-archetypical perspective on the foundations of human (and animal) subjectivity. Front Psychol 8.https://doi.org/10.3389/fpsyg.2017.01424
19. Bogerts B, Peter E, Schiltz K (2011) Aggression, Gewalt, Amok, Stalking. In: Möller HJ, Laux G, Kapfhammer HP (eds) Psychi. Berlin
20. Seidenbecher S, Steinmetz C, Möller-Leimkühler A-M, Bogerts B (2020) Terrorismus aus psychiatrischer Sicht. Nervenarzt 91(5):422–432. https://doi.org/10.1007/s00115-020-00894-0
21. Bogerts B, Möller-Leimkühler AM (2013) Neurobiological and psychosocial causes of individual male violence. Nervenarzt 84(11):1329–1344. https://doi.org/10.1007/s00115-012-3610-x
22. Pujara M, Motzkin JC, Newman JP, Kiehl KA, Koenigs M (2014) Neural correlates of reward and loss sensitivity in psychopathy. Soc Cogn Affect Neurosci. 9(6):794–801. https://doi.org/10.1093/scan/nst054
23. Jankowski KF, Takahashi H (2014) Cognitive neuroscience of social emotions and implications for psychopathology: examining embarrassment, guilt, envy, and schadenfreude. Psychiatry Clin Neurosci 68(5):319–336. https://doi.org/10.1111/pcn.12182
24. Lammers JHCM, Meelis W, Kruk MR, van der Poel AM (1987) Hypothalamic substrates for brain stimulation-induced grooming, digging and circling in the rat. Brain Res 418(1):1–19. https://doi.org/10.1016/0006-8993(87)90956-5
25. Wu Z, Autry AE, Bergan JF, Watabe-Uchida M, Dulac CG (2014) Galanin neurons in the medial preoptic area govern parental behaviour. Nature 509(7500):325–330. https://doi.org/10.1038/nature13307
26. Kosfeld M, Heinrichs M, Zak PJ, Fischbacher U, Fehr E (2005) Oxytocin increases trust in humans. Nature 435(7042):673–676. https://doi.org/10.1038/nature03701
27. Ross HE, Young LJ (2009) Oxytocin and the neural mechanisms regulating social cognition and affiliative behavior. Front Neuroendocrinol 30(4):534–547. https://doi.org/10.1016/j.yfrne.2009.05.004
28. Cservenák M, Keller D, Kis V et al (2017) A thalamo-hypothalamic pathway that activates oxytocin neurons in social contexts in female rats. Endocrinology 158(2):335–348. https://doi.org/10.1210/en.2016-1645
29. Bernhardt BC, Singer T (2012) The neural basis of empathy. Annu Rev Neurosci 35(1):1–23. https://doi.org/10.1146/annurev-neuro-062111-150536
30. Singer T, Klimecki OM (2014) Empathy and compassion. Curr Biol 24(18):R875–R878. https://doi.org/10.1016/j.cub.2014.06.054
31. Decety J, Skelly LR, Kiehl KA (2013) Brain response to empathy-eliciting scenarios involving pain in incarcerated individuals with psychopathy. JAMA Psychiat 70(6):638. https://doi.org/10.1001/jamapsychiatry.2013.27
32. Lockwood PL (2016) The anatomy of empathy: vicarious experience and disorders of social cognition. Behav Brain Res 311:255–266. https://doi.org/10.1016/j.bbr.2016.05.048
33. Decety J, Cowell JM (2014) Friends or Foes: Is Empathy Necessary for Moral Behavior? Perspect Psychol Sci 9(5):525–537. https://doi.org/10.1177/1745691614545130
34. Winter K, Spengler S, Bermpohl F, Singer T, Kanske P (2017) Social cognition in aggressive offenders: Impaired empathy, but intact theory of mind. Sci Rep 7(1):670. https://doi.org/10.1038/s41598-017-00745-0
35. Klimecki OM, Leiberg S, Lamm C, Singer T (2013) Functional neural plasticity and associated changes in positive affect after compassion training. Cereb Cortex 23(7):1552–1561. https://doi.org/10.1093/cercor/bhs142

Chapter 7
Brain Pathology in Violent Offenders

By studying the brain-biological correlates of aggression and those of compassion, it has become possible to understand why structural and functional disturbances of these brain regions are associated with an increased propensity to violence. This is supported by numerous studies of the brains of violent offenders.

Examining the Brain Using Imaging Techniques

Once it became possible to visualize in detail the structures within the brain using imaging techniques such as computer tomography and magnetic resonance imaging, a large number of brain examinations were carried out on violent offenders, beginning in the mid-1990s, in order to find out how they differed from the normal population. To date, there have been more than 100 such studies worldwide using brain structural or functional imaging techniques, most of which were performed on offenders in prisons who had antisocial personality disorder or psychopathy [1–15]. Antisocial personality offenders do not adhere to social rules, are reckless, impulsive, easily aroused, provocative and prone to reactive aggression. Psychopaths, on the other hand, are callous, uncompassionate, prone to violence to achieve their own advantage, but also use superficially winning and deceitful behavior and tend to be proactive, that is, their violence is planned, thought-out and deliberate (see Chap. 10). Although the imaging studies differ in some of their detailed findings, they agree that there are structural and functional deficits in brain regions responsible for moral behavior and in brain areas controlling aggression centers in the limbic system and brain stem. However, the vast majority of these studies report only statistical group differences, that is, differences in group means, although there may be a wide range of scatter in the results between individuals. Changes in brain structure or function need not be detectable in every offender.

In summary, the results of this substantial number of studies show that in the antisocial offender type with a propensity for reactive violence, there is a decreased volume of the anterior, middle, and inferior frontal brain, anterior and middle temporal brain, and the limbic central structures (amygdala and hippocampus) located therein (Fig. 7.1); also, increased reactivity of the amygdala to aggression-triggering stimuli. In proactively violent psychopaths, tissue deficits are present in the same anterior, inferior, and middle parts of the frontal brain, temporal brain, and limbic system; and additionally, in the insular cortex 11 (Fig. 7.1). The insular cortex is the most important brain region for empathic feeling and guilt consciousness. The amygdala is responsible for the emotional evaluation of the social environment. In psychopaths, the amygdala can only be activated to a diminished extent. This explains why these persons are more emotionless and fearless and therefore do not shrink from the consequences of their deeds and possible punishments, although they have no intelligence deficits. In functional brain images, the activity of the insula and the amygdala correlate inversely with the severity of psychopathy: the lower the activity of these brain regions important for moral action, the more pronounced the psychopathic character trait [14].

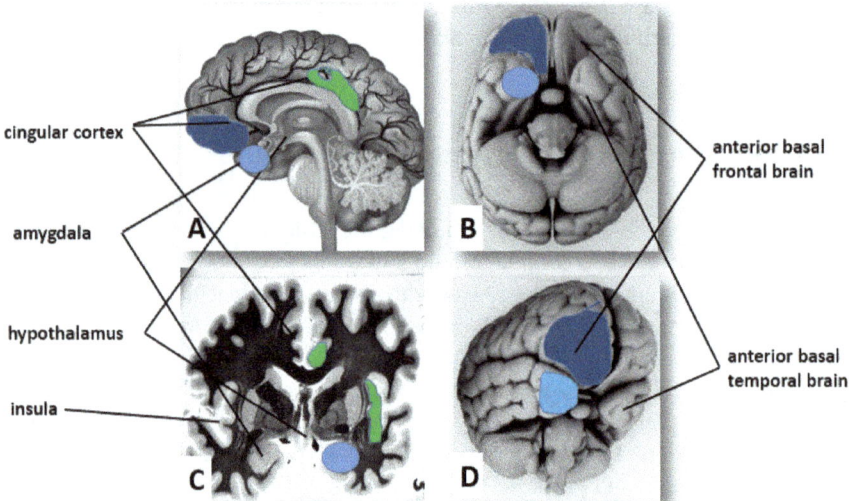

Fig. 7.1 Brain regions showing structural and functional deficits in violent offenders with antisocial personality disorder or psychopathy. Anterior, inferior and middle frontal brain (dark blue) is affected in both psychosyndromes; insular cortex and superior cingulate cortex (green) only in psychopathy. The amygdala (light blue) is hyperexcitable in antisocial disorder and smaller in psychopathy. **a** View of the brain surface located to the midline. **b** View from below. **c** Section through the center of the brain, view from the front, colored black are the nerve fiber tracts within the brain. **d** View from the lower right front. Own illustration. Partial images from Deposit photos No.: 86104948/12568185

Causes of Brain Structural and Functional Deficits

The question arises why in psychopaths and sociopaths these brain regions responsible for violence inhibition and compassionate attitudes show structural and functional deficits. Psychopathy is subject to a strong genetic influence. Up to 70% of the causal variance is inherited [16]. On the other hand, the brain is plastically formable by early social influences and gene activity is modifiable by epigenetic mechanisms (see Chap. 5). In other words, adequate early activation through positive emotional attention promotes growth and full functioning of these brain areas. If this support is lacking, prosocial traits develop insufficiently. The latter cause seems to apply predominantly to antisocial personality disorders. Thus, both genetic and neuroplastic (i.e., early social influences) factors play a causative role.

Historical Cases—Prominent Examples

Phineas Gage

The historically best-known case of personality change after injury to the brain structure is that of the railroad worker Phineas Gage. In 1848, while working on the railways, he stuffed explosives into a borehole with an iron rod, causing an explosion. The rod flew through his skull from below and pierced his left frontal brain. Gage survived this serious accident and made a full physical recovery after several weeks. He was, however, considerably changed in character. Before the accident he had a friendly and reserved nature; afterwards he was conspicuous for disinhibited and impulsive behavior; moral sense seemed to have been lost to him as a result of the accident. The iron bar had destroyed in the frontal brain a part of the brain regions whose intact function is necessary for the higher and more differentiated personality traits, including impulse and aggression control, and ethical-moral attitudes [6]. The case of Phineas Gage was the first report that triggered discussion of brain anatomical bases of moral or criminal behavior in the nascent neurosciences of that time.

Subsequently, numerous similar cases were reported. Individuals who had previously behaved in a completely inconspicuous manner had changes of personality with disinhibited, aggressive, and even criminal and violent behavior after brain damage due to injuries, tumors and infections, or after pathological degeneration of brain tissue in the anterior regions of the frontal and temporal brain (Figs. 6.8 and 7.1). Such patients are referred to as "pseudo-psychopathy" or "acquired sociopathy".

Head Schoolmaster Ernst Wagner

One of the most famous spree killers in the European history of psychiatry is the schoolteacher Wagner. In 1913, he killed his wife and four children in one night and then shot 9 people in the village of Mühlhausen near Stuttgart, Germany, seriously wounded 11 others and burned down several houses. He targeted exclusively the men of Mühlhausen. The fact that he also shot three girls and injured one woman was the only thing he later regretted.

Wagner was overpowered and made a mentally disturbed impression when he was arrested, which is why he was psychiatrically examined. The psychiatric expert diagnosed Wagner with delusional development accompanied by significant reality disturbance, which the expert considered to be the main cause of his rampage. Due to the identified psychiatric disorder, Wagner was classified as incapable of guilt, which is why he was admitted to a forensic psychiatric hospital. He remained in this hospital for many years until he died of tuberculosis. His brain was sent to the Vogt Brain Research Institute, where it was prepared in series of sections for neuropathological examination. The later examination of the brain showed that in the area of the left middle temporal brain there was a spatially very circumscribed, about one centimeter long, defect of the cortex next to the hippocampus and the amygdala, which was classified as a brain development disorder because of the distortions of the cerebral cortical layers that could be seen [11, 17]. This defect was thus located at a strategically important place of emotional evaluation and interpretation for everything perceived. Here, the impulses coming from the sensory fields of the cerebral cortex are transmitted to the limbic central structures, namely the amygdala and the hippocampus (Fig. 7.2).

In Wagner's brain, an important switching station for the adequate emotional classification and reality evaluation of the perceived environment was thus defective. This cortex defect in Wagner's limbic system is located at the same place where structural and functional deficits could be detected in some paranoid/hallucinatory patients in the context of a schizophrenic psychosis. It also corresponds to the area of the middle inferior temporal brain, which also shows structural and functional deficits in psychopaths and sociopaths in prison (Fig. 7.3). From the location of the lesion, both the emotional misinterpretation of the environment in the context of Wagner's delusional symptomatology and the deficient aggression control can be inferred.

Charles Whitman

In the weeks and months leading up to his spree killing, Texan Charles Whitman, a hitherto completely psychologically inconspicuous person, complained of what he saw as inexplicable emotional irritations, particularly a growing aggressiveness—until in 1966 he shot at anything that moved with a rifle from the University of Texas

Historical Cases—Prominent Examples

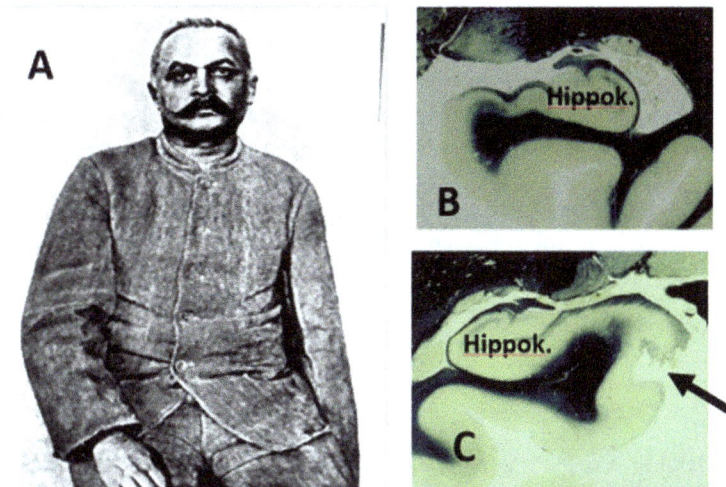

Fig. 7.2 **a** Head schoolmaster Ernst Wagner in 1913 as a forensic psychiatric patient in typical institutional clothing; **b** Normal hippocampal formation with normal adjacent cortex of the middle inferior temporal brain; **c** Brain slice preparation of the hippocampal formation of Wagner's brain with pathological structure (so-called invagination) of the adjacent temporal cortex (arrow). The region is located at the site of the temporal brain also marked in Fig. 7.3. **a** Head teacher Ernst Wagner. Photo taken in 1913 (public domain). **b, c** own photos

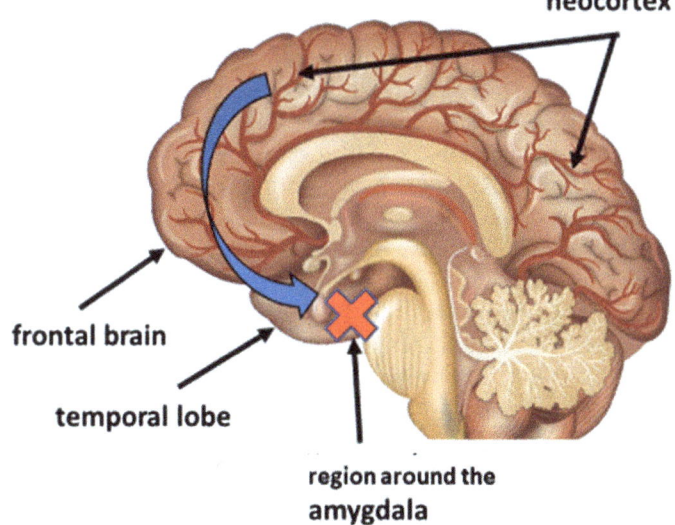

Fig. 7.3 Midline view of the brain. The location damaged in the cases of Wagner, Meinhof, and Whitman is marked X. The curved arrow indicates that information preprocessed in the cortex converges at this site to be transmitted to the amygdala for emotional evaluation. Own illustration. Partial images from Deposit photos No.: 86104948

tower, killing 17 people and injuring 32 others before being shot dead by police. He also killed the unborn child of a pregnant survivor. By morning, Whitman had already stabbed his mother and wife to death. Before committing his crimes, he wrote a testament requesting that after his death his brain should be examined, because he felt that something had changed in it. The autopsy of his brain revealed a tumor the size of a walnut, located next to the right amygdala [18].

Similar to Wagner and Meinhof (see below), this tumor caused a local (regionally circumscribed) structural and functional impairment in a limbic part of the brain that is of central importance for emotion and aggression control.

Ulrike Meinhof

Ulrike Meinhof was not a person who ran amok and therefore her actions are not comparable to the aforementioned examples. She was the leading intellectual figure of left-wing terrorism in Germany during the 1970s. Before joining the terrorist group Red Army Faction (RAF) in 1970, she was a widely recognized journalist who used peaceful means and remarkable verbal ability to advocate the political goals she represented. In 1962, after the onset of neurological symptoms, she underwent brain surgery at the Neurosurgical University Clinic in Hamburg. During the operation, an attempt was made to surgically remove a vascular tumor that had previously been identified by X-ray and was located at the base of the brain. In the years that followed, there was a personality change with growing aggressive/violent traits that were foreign to her former nature. In 1971 she wrote the *Concept of the Urban Guerrilla*, in which she argued that the guerrilla strategies of Central and South American guerrilla fighters should also be transferred to German cities, for the overthrow of the ruling capitalist system. She committed suicide in 1976 during the trial of the leading members of the RAF in the high-security prison in Stuttgart/Stammheim. In a neuropathological report drawn up on the basis of an autopsy ordered by the public prosecutor's office, extensive damage was found in the middle parts of the right temporal brain, especially in an area of the cerebral cortex lying next to the amygdala, which was attributable to the 1962 brain operation (Fig. 7.3).

The location of the damage in a limbic brain area important for aggression control was considered to be the neuropathological foundation for the personality change with increased aggression and violence potential that occurred in the years after her surgery [11, 19, 20]. However, brain pathology only explains the pathological emotional component, not the actual contents of the thoughts; these are to be categorized against the background of the specific psychosocial and political environment of the time.

Brain Pathology of Imprisoned Violent Offenders

The brain pathological findings in the violent offenders described are not isolated cases that would not be representative of a significant percentage of aggressive perpetrators. Similar brain pathologies are to be found in a substantial number of individuals convicted of serious violent crimes and imprisoned.

Our group evaluated a large number of magnetic resonance imaging and computed tomography images of the brain structure of violent offenders in a big German penitentiary, which the prison physician had ordered to exclude brain disease in individuals reporting dizziness or headaches. These images were compared with brain images of prisoners convicted of offenses other than violent crimes, such as fraud or drug trafficking, and with brain structure images of persons who had not committed any crime [21].

The brain images were evaluated without knowledge of the type of offense or whether the particular brain belonged to a non-offending or a convicted person. Only after the evaluations were completed was the allocation to the respective group made for statistical evaluation. It was found that especially the frontal brain, both right and left, and the temporal brain, here again in brain areas adjacent to the central limbic structures, showed a significantly higher rate of pathological brain tissue changes in the violent offenders than in the non-violent convicts and in the non-offenders. Assessments of magnetic resonance imaging and computed tomography scans, as is common in routine clinical practice, detected brain tissue changes or structural deficits in approximately one-third of the violent offenders (Fig. 7.4) [21].

In the case of the perpetrators who showed brain structural abnormalities, this was not an issue raised during the court hearings, probably because it was not known and such persons are usually inconspicuous in their everyday behavior. Only above a

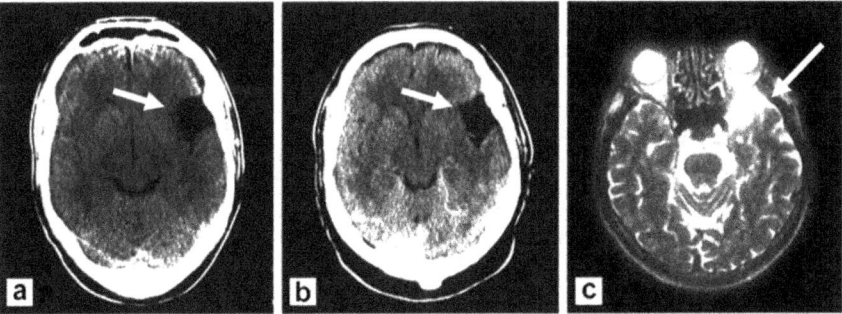

Fig. 7.4 a, b Computed tomographic images of the heads of two convicted violent offenders (skull bone is light, brain tissue is gray, cerebrospinal fluid is dark). The dark areas in the right temporal brain (arrow) are dilated cerebrospinal fluid spaces with significant volume reduction of the surrounding brain tissue. This area belongs to the brain regions responsible for moral behavior and aggression control, among others (compare Fig. 7.1). **c** Magnetic resonance image of another violent offender (cerebrospinal fluid bright) with a similar tissue defect in the right anterior temporal brain. Own photos

certain stress threshold, for example, where there has been provocation, is the control of aggressive impulses disturbed in such perpetrators, because the control regions of the brain responsible for this are defective.

Comparable findings were also obtained in several studies of American prison inmates convicted of violent crimes [3–5, 15, 22].

References

1. Raine A (2013) The anatomy of violence—the biological roots of crime; Chapter 1. Penguin Books, London
2. Glenn AL, Raine A, Schug RA (2009) The neural correlates of moral decision-making in psychopathy. Mol Psychiatry 14(1):5–6. https://doi.org/10.1038/mp.2008.104
3. Kiehl KA, Smith AM, Hare RD et al (2001) Limbic abnormalities in affective processing by criminal psychopaths as revealed by functional magnetic resonance imaging. Biol Psychiatry 50(9):677–684. http://www.ncbi.nlm.nih.gov/pubmed/11704074
4. Cope LM, Ermer E, Gaudet LM et al (2014) Abnormal brain structure in youth who commit homicide. NeuroImage Clin. 4:800–807. https://doi.org/10.1016/j.nicl.2014.05.002
5. Ermer E, Cope LM, Nyalakanti PK, Calhoun VD, Kiehl KA (2013) Aberrant paralimbic gray matter in incarcerated male adolescents with psychopathic traits. J Am Acad Child Adolesc Psychiatry 52(1):94-103.e3. https://doi.org/10.1016/j.jaac.2012.10.013
6. Damasio H, Grabowski T, Frank R, Galaburda A, Damasio A (1994) The return of Phineas Gage: clues about the brain from the skull of a famous patient. Science (80-), 264(5162):1102–1105. https://doi.org/10.1126/science.8178168
7. Damasio AR, Tranel D, Damasio H (1990) Individuals with sociopathic behavior caused by frontal damage fail to respond autonomically to social stimuli. Behav Brain Res 41(2):81–94. http://www.ncbi.nlm.nih.gov/pubmed/2288668
8. Raine A, Buchsbaum M, Lacasse L (1997) Brain abnormalities in murderers indicated by positron emission tomography. Biol Psychiatry 42(6):495–508. https://doi.org/10.1016/S0006-3223(96)00362-9
9. Raine A, Yang Y (2006) Neural foundations to moral reasoning and antisocial behavior. Soc Cogn Affect Neurosci. 1(3):203–213. https://doi.org/10.1093/scan/nsl033
10. Witzel JG, Bogerts B, Schiltz K (2016) Increased frequency of brain pathology in inmates of a high-security forensic institution: a qualitative CT and MRI scan study. Eur Arch Psychiatry Clin Neurosci 266(6):533–541. https://doi.org/10.1007/s00406-015-0620-2
11. Bogerts B, Schöne M, Breitschuh S (2018) Brain alterations potentially associated with aggression and terrorism. CNS Spectr 23(02):129–140. https://doi.org/10.1017/S1092852917000463
12. Poeppl TB, Donges MR, Mokros A et al (2019) A view behind the mask of sanity: meta-analysis of aberrant brain activity in psychopaths. Mol Psychiatry 24(3):463–470. https://doi.org/10.1038/s41380-018-0122-5
13. Darby RR, Horn A, Cushman F, Fox MD (2018) Lesion network localization of criminal behavior. Proc Natl Acad Sci 115(3):601–606. https://doi.org/10.1073/pnas.1706587115
14. Johanson M, Vaurio O, Tiihonen J, Lähteenvuo M (2020) A systematic literature review of neuroimaging of psychopathic traits. Front Psychiatry 10. https://doi.org/10.3389/fpsyt.2019.01027
15. Anderson NE, Kiehl KA (2013) Psychopathy and aggression: when paralimbic dysfunction leads to violence, pp 369–393. https://doi.org/10.1007/7854_2013_257
16. Bezdjian S, Raine A, Baker LA, Lynam DR (2011) Psychopathic personality in children: genetic and environmental contributions. Psychol Med 41(03):589–600. https://doi.org/10.1017/S0033291710000966

References

17. Bogerts B (2006) Gehirn und Verbrechen: Neurobiologie von Gewalttaten. In: Schneider F (ed) Entwicklungen in Der Psychiatrie. Springer, Heidelberg, pp 335–347. https://link.springer.com/chapter/10.1007/3-540-30100-3_35
18. Eagleman D The brain on trial. The Atlantic. https://www.theatlantic.com/magazine/archive/2011/07/the-brain-on-trial/308520/. Published 2011. Accessed August 26, 2020
19. Finn P Germans studied brains of radical group's leaders. Washington Post. https://www.washingtonpost.com/archive/politics/2002/11/19/germans-studied-brains-of-radical-groups-leaders/7259f0a2-5b70-43fc-99dc-14f00e0e7bf6/. Published 2002
20. BBC Meinhof brain study yields clues. news.bbc.co.uk. http://news.bbc.co.uk/2/hi/europe/2455647.stm. Published 2002
21. Schiltz K, Witzel JG, Bausch-Hölterhoff J, Bogerts B (2013) High prevalence of brain pathology in violent prisoners: a qualitative CT and MRI scan study. Eur Arch Psychiatry Clin Neurosci 263(7):607–616. https://doi.org/10.1007/s00406-013-0403-6
22. Cope LM, Ermer E, Nyalakanti PK, Calhoun VD, Kiehl KA (2014) Paralimbic gray matter reductions in incarcerated adolescent females with psychopathic traits. J Abnorm Child Psychol 42(4):659–668. https://doi.org/10.1007/s10802-013-9810-4

Chapter 8
The Role of Hormones and Messenger Substances in the Brain

From a neurobiological point of view, in addition to pathological deviations in brain anatomy and physiology, alterations in brain chemistry are also relevant for violent behavior. This is influenced by a large number of hormones and neuronal transmitter substances. The most important of these are testosterone, oxytocin, cortisol and serotonin.

Testosterone

Since violent acts are committed more often by men than by women, it is tempting to assume that the male sex hormone testosterone is partly responsible for this.

In most mammals, a correlation between testosterone levels and aggressiveness has indeed been found. In humans, however, the findings are not so clear. What seems to be certain is the connection between the level of testosterone in the blood and social dominance in males, less so with aggressive behavior [1, 2]. When a rodent emerges victorious from its fight for territory in laboratory experiments, the number of testosterone receptors in the brain's reward system increases, amplifying the effects of testosterone. The connection between testosterone levels and male social dominance is also evident in the fact that after sporting competitions, testosterone levels rise in the winners—interestingly, however, not only in the players, but also in the supporters of the winning team. Thus, the feeling of belonging to the winning group is sufficient for increased blood levels in this hormone [2]. The mass euphoria demonstrated by sports fans after championship wins may be related to this.

Some researchers have found higher testosterone levels in violent offenders in prison than in non-violent prisoners. A correlation between testosterone levels and propensity to violence in young adulthood has also been reported [3, 4]. Testosterone levels were significantly higher in a group of young violent prisoners than in those jailed for other offenses [5]. The use of anabolic steroids, which are substances with testosterone-like effects often taken by bodybuilders, has been reported to be associated with higher propensity to violence [4].

Because of the rather low correlation between testosterone levels and inclination to violence, it seems unlikely that the male Y chromosome explains the higher violent crime rate of the "stronger gender" only via this sex hormone. More important, on the other hand, seem to be gender differences in the fine-tissue brain structure in emotion-relevant brain areas, which are located in the deep structures of the brain stem and the limbic system and are responsible not only for different sexual behavior between men and women, but also for the emergence and control of aggression (see Chap. 9).

Oxytocin

Oxytocin is produced in cell groups of the hypothalamus and not only enters the bloodstream as a hormone, but also influences other brain regions in the brain stem and limbic system that are important for emotions and feelings. It plays a significant role in the birth process and in mother–child bonding. It is also known as the feel-good and bonding hormone, as it not only promotes bonding between parents and children, but is also activated in couples and in interpersonal closeness in general as well as in sympathy. Oxytocin makes people more generous; this hormone is also released in recipients of prosocial behavior [6]. Oxytocin administered as a nasal spray promotes positive communication, and reduces anxiety and the release of the stress hormone cortisone [7]. Oxytocin diminishes aggressiveness in rodents. Mice that had their oxytocin system turned off (by eliminating the gene encoding the oxytocin receptor) were extremely aggressive [8].

On the other hand, the anti-aggressive effects of oxytocin are not quite so unequivocal. Certain variants of oxytocin receptors in the brain are associated with antisocial behavior [9]. It produces benevolence and cooperativeness only toward acquaintances and friends, and promotes rejection of strangers [10]. Attempts to use it as an anti-aggressive drug have yielded inconsistent results: in men, both positive and absent effects have been reported [11, 12]. In women, oxytocin appears to have aggression-increasing effects toward strangers to protect their offspring [13].

Stress Hormone Cortisol

In stressful situations, the hormone cortisol is released from the adrenal cortex. This not only affects the heart, circulation, metabolism and immune system, but also the brain, which has a high density of cortisol receptors in the limbic system and hypothalamus.

Permanent stress makes people not only depressed, but also aggressive. This is due to the fact that stress-induced elevated cortisol concentrations impair the function of the frontal cortex and thus the ability to plan ahead, to act in a deliberate manner, to assess risks and to control impulses. During prolonged stress, hectic actions, irritability and resulting reactive aggressiveness become particularly apparent. The connection of signals from the prefrontal cortex to the amygdala is impaired by the stress hormone, and with it the ability of the frontal brain to control the areas of the brainstem relevant to aggression [14, 15].

When stress-induced cortisol floods the brain for prolonged periods, it damages brain cells by limiting the action of neuronal growth factors. This is particularly harmful in cases of early childhood trauma and its later sequelae, such as borderline disorder (see Chap. 10). In these patients, shrinkage of the hippocampus, amygdala and related cortical areas is found as a result of childhood stressors, which adversely affects cognition, impulse control and mental balance, and thus increases irritability and aggressiveness [14, 16–18]. Stress reduction (if at all possible) is therefore an effective strategy for preventing aggression.

Serotonin

Serotonin is one of the neuronal messenger substances necessary for the regulation of mental state and emotional reactions, including aggression. Decreased activity of serotonin, sometimes called the "happiness hormone" because of its positive effect on the mind, correlates with aggressive behavior in animal studies. A deficiency of serotonin leads to low mood, increased irritability and higher propensity to violence [19–21]. This is not surprising, since a decrease in serotonin has also been found in the brains of patients who died by suicide. Psychodynamic similarities exist between self-aggression, of which suicide is the most dramatic form, and aggression towards others; serotonin is decreased in the brain in both forms of aggression [22].

If blood levels of serotonin are lowered by reducing dietary tryptophan, an amino acid essential for the synthesis of serotonin, aggressiveness increases. If the activity of serotonin at the neuronal contact points in the brain (synapses) is improved again by a special group of antidepressants irritability and aggressiveness is reduced. However, this is true only for impulsive-reactive aggression, not for planned, proactive aggression [20, 23–25]. Psychotropic drugs that improve the action of serotonin in the brain are not only antidepressants, but also have anti-aggressive properties.

In contrast, there have been sporadic reports that treatment with certain antidepressants that enhance serotonin action in the brain has also resulted in increased acts of violence. In a recent very large study conducted on almost 800,000 people, it was shown that this was true only for a very small minority of patients (2.6%) who were treated with such antidepressants. 97% of those treated did not become violent compared to 98% of the general population [26].

References

1. Archer J (2006) Testosterone and human aggression: an evaluation of the challenge hypothesis. Neurosci Biobehav Rev 30(3):319–345. https://doi.org/10.1016/j.neubiorev.2004.12.007
2. Carré JM, Archer J (2018) Testosterone and human behavior: the role of individual and contextual variables. Curr Opin Psychol 19:149–153. https://doi.org/10.1016/j.copsyc.2017.03.021
3. O'Connor DB, Archer J, Wu FCW (2004) Effects of testosterone on mood, aggression, and sexual behavior in young men: a double-blind, placebo-controlled, cross-over study. J Clin Endocrinol Metab 89(6):2837–2845. https://doi.org/10.1210/jc.2003-031354
4. Giotakos O, Markianos M, Vaidakis N, Christodoulou GN (2004) Sex hormones and biogenic amine turnover of sex offenders in relation to their temperament and character dimensions. Psychiatry Res 127(3):185–193. https://doi.org/10.1016/j.psychres.2003.06.003
5. Dabbs JM, Frady RL, Carr TS, Besch NF (1987) Saliva testosterone and criminal violence in young adult prison inmates. *Psychosom Med*. 49(2):174–182. http://www.ncbi.nlm.nih.gov/pubmed/3575604.
6. Zak P, Kurzban R, Matzner W (2005) Oxytocin is associated with human trustworthiness. Horm Behav 48(5):522–527. https://doi.org/10.1016/j.yhbeh.2005.07.009
7. Ditzen B, Schaer M, Gabriel B, Bodenmann G, Ehlert U, Heinrichs M (2009) Intranasal oxytocin increases positive communication and reduces cortisol levels during couple conflict. Biol Psychiatry 65(9):728–731. https://doi.org/10.1016/j.biopsych.2008.10.011
8. Dhakar MB, Rich ME, Reno EL, Lee H-J, Caldwell HK (2012) Heightened aggressive behavior in mice with lifelong versus postweaning knockout of the oxytocin receptor. Horm Behav 62(1):86–92. https://doi.org/10.1016/j.yhbeh.2012.05.007
9. Hovey D, Lindstedt M, Zettergren A et al (2016) Antisocial behavior and polymorphisms in the oxytocin receptor gene: findings in two independent samples. Mol Psychiatry 21(7):983–988. https://doi.org/10.1038/mp.2015.144
10. Declerck CH, Boone C, Kiyonari T (2010) Oxytocin and cooperation under conditions of uncertainty: the modulating role of incentives and social information. Horm Behav 57(3):368–374. https://doi.org/10.1016/j.yhbeh.2010.01.006
11. Berends YR, Tulen JHM, Wierdsma AI et al (2019) Intranasal administration of oxytocin decreases task-related aggressive responses in healthy young males. Psychoneuroendocrinology 106:147–154. https://doi.org/10.1016/j.psyneuen.2019.03.027
12. Alcorn JL, Green CE, Schmitz J, Lane SD (2015) Effects of oxytocin on aggressive responding in healthy adult men. Behav Pharmacol 26(8 Spec No):798–804. https://doi.org/10.1097/FBP.0000000000000173
13. de Jong TR, Neumann ID (2017) Oxytocin and aggression. In: Hurlemann R, Grinevich V (eds) Behavioral Pharmacology of Neuropeptides: Oxytocin. Current Topics in Behavioral Neurosciences, vol 35. Springer, Cham. https://doi.org/10.1007/7854_2017_13
14. Young AH, Sahakian BJ, Robbins TW, Cowen PJ (1999) The effects of chronic administration of hydrocortisone on cognitive function in normal male volunteers. Psychopharmacology 145(3):260–266. https://doi.org/10.1007/s002130051057
15. Barsegyan A, Mackenzie SM, Kurose BD, McGaugh JL, Roozendaal B (2010) Glucocorticoids in the prefrontal cortex enhance memory consolidation and impair working memory by a common neural mechanism. Proc Natl Acad Sci 107(38):16655–16660. https://doi.org/10.1073/pnas.1011975107
16. Sandi C, Haller J (2015) Stress and the social brain: behavioural effects and neurobiological mechanisms. Nat Rev Neurosci 16(5):290–304. https://doi.org/10.1038/nrn3918
17. Sapolsky RM (1996) Why stress is bad for your brain. Science (80-) 273(5276):749–750. https://doi.org/10.1126/science.273.5276.749
18. Herpertz SC, Nagy K, Ueltzhöffer K et al (2017) Brain mechanisms underlying reactive aggression in borderline personality disorder-sex matters. Biol Psychiatry 82(4):257–266. https://doi.org/10.1016/j.biopsych.2017.02.1175

References

19. Coccaro EF, Fanning JR, Phan KL, Lee R (2015) Serotonin and impulsive aggression. CNS Spectr 20(03):295–302. https://doi.org/10.1017/S1092852915000310
20. Duke AA, Bègue L, Bell R, Eisenlohr-Moul T (2013) Revisiting the serotonin–aggression relation in humans: a meta-analysis. Psychol Bull 139(5):1148–1172. https://doi.org/10.1037/a0031544
21. Moffitt TE, Brammer GL, Caspi A et al (1998) Whole blood serotonin relates to violence in an epidemiological study. Biol Psychiatry 43(6):446–457. https://doi.org/10.1016/S0006-3223(97)00340-5
22. Virkkunen M, Goldman D, Nielsen DA, Linnoila M (1995) Low brain serotonin turnover rate (low CSF 5-HIAA) and impulsive violence. J Psychiatry Neurosci 20(4):271–275. http://www.ncbi.nlm.nih.gov/pubmed/7544158
23. Dougherty DM, Moeller FG, Bjork JM, Marsh DM (1999) Plasma L-tryptophan depletion and aggression. Adv Exp Med Biol 467:57–65. https://doi.org/10.1007/978-1-4615-4709-9_7
24. Vartiainen H, Tiihonen J, Putkonen A, et al (1995) Citalopram, a selective serotonin reuptake inhibitor, in the treatment of aggression in schizophrenia. Acta Psychiatr Scand 91(5):348–351. http://www.ncbi.nlm.nih.gov/pubmed/7639092
25. Berman ME, McCloskey MS, Fanning JR, Schumacher JA, Coccaro EF (2009) Serotonin augmentation reduces response to attack in aggressive individuals. Psychol Sci 20(6):714–720. https://doi.org/10.1111/j.1467-9280.2009.02355.x
26. Lagerberg T, Fazel S, Molero Y et al (2020) Associations between selective serotonin reuptake inhibitors and violent crime in adolescents, young, and older adults—a Swedish register-based study. Eur Neuropsychopharmacol 36:1–9. https://doi.org/10.1016/j.euroneuro.2020.03.024

Chapter 9
Gender Difference in the Propensity for Violence

The use of physical violence is a male domain; 90% of serious acts of violence are committed by men (see Chap. 3, Figs. 3.3 and 3.4). This applies to all forms of individual as well as collective violence. Why?

Phylogenetic Causes

Already in infancy, boys are more physically aggressive than girls [1]. This indicates that it is not so much learned psychosocial role behavior as gender differences in brain function predetermined (i.e., genetically based), or more precisely, formed during phylogenetic development, that are responsible for the higher male propensity to violence [2]. These differences are found not only in humans, but in almost all vertebrates.

The fact that females are less inclined to use physical violence in provocative situations and are overall more risk averse than males, whereas the latter are more swashbuckling and more prone to fisticuffs, is as much the result of a long phylogenetic selection as of the gender differences in physical constitution. The care of children and the survival chances of the offspring will be less good with a mother who is willing to behave violently and to take risks. A risk-averse and non-combative male, on the other hand, has less chance of out-competing his male rivals and of passing on his genes by gaining a dominant position. The test of strength, the imposition of one's own power, threats and intimidation of rivals are a male domain in all mammals including *Homo sapiens*. These mental gender differences, which are already detectable in childhood, are the result of the genetically Y-chromosomally predetermined, thus evolutionarily formed, brain structural and functional differences between males and females. The level of testosterone in the blood plays only a secondary role [3].

Brain Biological Correlates of Sex Difference

Male and female brains are not the same. Numerous comparisons of male and female brains using computed tomographic and magnetic resonance imaging methods have demonstrated subtle but statistically significant sex differences in the structure and function of several brain regions [4, 5]. For example, brain functional imaging techniques (functional nuclear magnetic resonance imaging) have demonstrated that the reward system responds more to monetary incentives in the male brain and more to social affirmation in the female brain [6]. The violence-regulating centers in the brain also exhibit gender differences. Men have a larger amygdala and higher amygdala activation during provocation than women [7]. In contrast, parts of the frontal brain responsible for deliberative behavior are larger in women. Inhibition of the amygdala, which becomes active when provoked, by cortical areas of the middle frontal brain (cingulate cortex) works better in women than in men [8, 9]. Therefore, women are more restrained in such situations.

The most pronounced difference in brain tissue size between the sexes is found in a group of cells in a region of the anterior hypothalamus responsible for sexual behavior (sexually dimorphic nucleus). This area is two to three times larger in men than in women. The cell group is located in close proximity to the aggression-relevant as well as the oxytocin-containing areas [10]. In the anterior hypothalamus and in the amygdala, the highest density of testosterone receptors is found [11]. This makes the connection between male aggression, especially in the mating season in almost all mammals, and the accompanying male aggression to achieve or defend an alpha position, understandable. In this position, the chance to pass on one's genes and the character traits that depend on them is greatest.

An interesting physical sex difference that also says something about aggressiveness is the length ratio between the index and ring fingers, which is controlled by prenatal testosterone levels. Males have a longer ring finger relative to the index finger than females. Higher testosterone levels are associated with a longer ring finger. The length of the ring finger correlates with male character traits such as more pronounced dominance striving, risky behavior, aggressiveness and tendency toward violence [12, 13].

Instead of physical violence, women tend to use softer forms of relational aggression such as bullying, exclusion and defamation. When direct physical and indirect relational aggression are considered together, the level of aggressiveness of both genders is roughly equal [2].

References

1. Potegal M, Archer J (2004) Sex differences in childhood anger and aggression. Child Adolesc Psychiatr Clin N Am 13(3):513–528. https://doi.org/10.1016/j.chc.2004.02.004
2. Björkqvist K (2018) Gender differences in aggression. Curr Opin Psychol 19:39–42. https://doi.org/10.1016/j.copsyc.2017.03.030

3. Carré JM, Archer J (2018) Testosterone and human behavior: the role of individual and contextual variables. Curr Opin Psychol 19:149–153. https://doi.org/10.1016/j.copsyc.2017.03.021
4. Pallayova M, Brandeburova A, Tokarova D (2019) Update on sexual dimorphism in brain structure-function interrelationships: a literature review. Appl Psychophysiol Biofeedback 44(4):271–284. https://doi.org/10.1007/s10484-019-09443-1
5. de Lacy N, McCauley E, Kutz JN, Calhoun VD (2019) Multilevel mapping of sexual dimorphism in intrinsic functional brain networks. Front Neurosci 13. https://doi.org/10.3389/fnins.2019.00332
6. Spreckelmeyer KN, Krach S, Kohls G et al (2009) Anticipation of monetary and social reward differently activates mesolimbic brain structures in men and women. Soc Cogn Affect Neurosci 4(2):158–165. https://doi.org/10.1093/scan/nsn051
7. Repple J, Habel U, Wagels L, Pawliczek CM, Schneider F, Kohn N (2018) Sex differences in the neural correlates of aggression. Brain Struct Funct 223(9):4115–4124. https://doi.org/10.1007/s00429-018-1739-5
8. Ruigrok ANV, Salimi-Khorshidi G, Lai M-C et al (2014) A meta-analysis of sex differences in human brain structure. Neurosci Biobehav Rev 39:34–50. https://doi.org/10.1016/j.neubiorev.2013.12.004
9. Raine A, Yang Y, Narr KL, Toga AW (2011) Sex differences in orbitofrontal gray as a partial explanation for sex differences in antisocial personality. Mol Psychiatry 16(2):227–236. https://doi.org/10.1038/mp.2009.136
10. Swaab D, Fliers E (1985) A sexually dimorphic nucleus in the human brain. Science (80-) 228(4703):1112–1115. https://doi.org/10.1126/science.3992248
11. Patchev VK, Schroeder J, Goetz F, Rohde W, Patchev AV (2004) Neurotropic action of androgens: principles, mechanisms and novel targets. Exp Gerontol 39(11–12):1651–1660. https://doi.org/10.1016/j.exger.2004.07.011
12. Stenstrom E, Saad G, Nepomuceno MV, Mendenhall Z (2011) Testosterone and domain-specific risk: digit ratios (2D:4D and rel2) as predictors of recreational, financial, and social risk-taking behaviors. Pers Individ Dif 51(4):412–416. https://doi.org/10.1016/j.paid.2010.07.003
13. Bailey AA, Hurd PL (2005) Finger length ratio (2D:4D) correlates with physical aggression in men but not in women. Biol Psychol 68(3):215–222. https://doi.org/10.1016/j.biopsycho.2004.05.001

Chapter 10
Mental Disorders and Violence

Events such as the deliberate crash of the Germanwings plane in March 2015, and the rampages in Oslo/Utøya Island in July 2011, in Las Vegas in October 2017, and Hanau, Germany, in February 2020, whose perpetrators are said to have been mentally ill (see Chaps. 16 and 17), gave rise to the impression that people with psychiatric disorders are sometimes prone to particularly serious acts of violence. In fact, only a very small minority of mentally ill people become violent. The vast majority of people with serious mental disorders do not commit criminal acts. How high is the actual risk of violence posed by mentally disturbed people?

General Risk of Violence in Mental Disorders

Several psychiatric disorders are associated with a slightly increased risk of violence. However, it must be emphasized that this risk exists only in a small proportion of such patients, even though it is more common in certain mental disorders than in the average population [1].

A large study of over 34,000 people published in 2009 found that 95% of people with severe mental illness (which included schizophrenia, bipolar disorder, major depression, and alcohol and drug addiction) lived violence-free lives compared to 98% of people without such illnesses [2, 3]. When interpersonal violence was perpetrated for the first time, mental disorders could explain the acts in 10% of men and 26% of women [4]. The risk is significantly increased when drug or alcohol use occurs alongside the mental illness.

In a British study conducted over a 10-year period, approximately 11% of violent offenders were diagnosed with a mental disorder. The same percentage was found for homicides. The most common diagnoses were addictive disorders, schizophrenia and other delusional disorders [5]. A Swedish study, however, found much higher numbers for the most serious form of violent crime, homicide; 90% of those convicted of murder or manslaughter were diagnosed with a mental disorder, the majority suffering from personality disorders, followed by schizophrenic psychosis and substance abuse [6].

The recidivism risk for incarcerated violent offenders on their release also depends on whether a mental illness is present. Five years after release, approximately 42% of former male prisoners with a psychiatric diagnosis and approximately 27% without such a diagnosis had engaged in interpersonal violence [7]. Psychiatric treatment after release significantly reduces the risk of future violent offenses. Proper drug treatment of offenders who present with psychosis halves the risk of future violence [8].

The frequency of violent acts varies greatly for individual diagnostic groups; various authors give the percentage of mentally disturbed persons who become violent in the course of their lives as in the range of 4–12%, depending on the type of data collection and definition of the illness. The greatest risk of violence is posed by alcohol and drug users and by persons with brain damage (organic brain disorders), as well as by patients diagnosed with schizophrenic diseases.

A recent, very extensive Swedish study of several hundred thousand people came to the conclusion that over a 10-year period and taking all diagnoses together, 6–7% of mentally ill patients became violent compared to 0.6–0.9% of mentally healthy people. However, the patients' risk of becoming victims of a violent act themselves was also just as high. Mentally ill patients were thus about seven times more likely than healthy people to be both perpetrators and victims of violence [3]. However, if mentally ill patients were compared not with the general population but with their mentally healthy relatives who came from the same social environment, the risk of violence was only four to five times higher. This was explained by the fact that, in addition to illness factors, the close social environment is also partly responsible for the occurrence of violence and exerts the same influences on the behavior of both patients and relatives [3].

The same study determined the different risk of violence for individual disease groups. For every 1000 persons affected by the respective disease, the following frequencies were given for the period of one year. The greatest risk was for drug addicts (approx. 25 out of 1000 addicts became violent per year), followed by alcohol addiction (approx. 13), personality disorders (approx. 12), schizophrenia (approx. 12), bipolar disorder (approx. 5), anxiety disorders (approx. 5) depression (approx. 4). In those with no psychiatric disorder the frequency was approx. 0.8/1000/year [3].

Schizophrenic and Psychotic Diseases

The typical psychopathological features of psychoses are disturbances of reality in the form of delusions (feelings of threat and persecution, of interference or of being manipulated) and auditory hallucinations (e.g., hearing commanding or commenting voices) or, more rarely, visual hallicinations. Psychotic diseases can have various causes such as drugs, brain inflammation, brain tumors or brain atrophy. The most common causes are schizophrenic disorders, which usually begin in young adulthood after months or even years of precursory stages. About 1% of the population develops schizophrenia during their lifetime. The vast majority of people diagnosed with this disorder behave peacefully, although they are often very withdrawn and suspicious, despite the considerable mental impairments, such as sensory illusions and disturbances of real perception. However, several extensive studies have demonstrated that there is a statistically increased risk of violent acts by affected individuals, especially when drug or alcohol use is added [1, 9].

The cause of the increased tendency of those with psychotic symptoms to violence is delusional perceptions and sensations of threat, persecution, defamation or interference, leading those affected to believe they have to defend themselves against perceived persecutors or attackers. This increased risk is particularly true for untreated patients at the onset of the disorder in young adulthood. Severe violence was found in 17% of first-time psychotic patients [10]. The risk was particularly high in patients with a history of previous violence and concomitant drug use [2]. Not only violence against others, but also aggression against oneself in the form of suicidal acts is a real risk in such patients [11].

According to an American study, subthreshold psychotic symptoms, so-called subclinical forms of schizophrenia without full expression of delusions and hallucinations, are present in about 5% of the general population, often in advance of full-blown psychotic symptoms becoming apparent later. In such persons the relative risk of violence, including the killing of other persons, was almost as high as in fully developed forms of schizophrenia [12]. As early as the first half of the last century there were impressive descriptions of murders, assassinations and rampages in the start-up phase (prodromal stage) of schizophrenia [13]. This phenomenon is therefore by no means new.

Death threats by schizophrenic patients towards relatives or therapeutic staff are to be taken seriously, since the possibility of the realization of such an act cannot be excluded, especially in case of insufficient treatment and deficient therapeutic vigilance.

With timely and adequate treatment with effective therapeutic measures, in which medications (so-called neuroleptics) also play an important role, the risk of violence and suicide is significantly reduced. Therefore, it is important to recognize the development of a schizophrenic disorder in time. The initial phase of such illnesses, which can often last for many months in young adulthood, is not always easy to recognize diagnostically and requires professional expertise. If a young person suddenly appears changed in character, without any other cause being apparent, behaves strangely, withdraws more and more, and is also preoccupied with occult matters, a psychiatrist should be consulted, who may be able to make a diagnosis of emerging psychosis and initiate preventive measures. Many larger psychiatric facilities have early detection units with staff trained for early diagnosis.

Depressive Disorders

The risk of violence due to depressive disorders is significantly lower than that from psychotic syndromes. According to a Swedish study, 3% of depressed individuals are at risk of violence, and one in 20 acts of violence was attributed to depression [14]. In the rare cases where depressives become violent, it is not so much the depressive illness itself as additional emerging mental health problems, such as substance use, personality disorders (especially borderline personality disorder) and experiences of violence in one's childhood, that contribute to aggressiveness. The diagnosis of depression itself contributes to the increased risk of violence mainly through its coincidence with such additional complications. Depression is less associated with overt than with covert aggression.

Statistically confirmed is a higher risk of interpersonal violence and homicide in suicidal persons, in which aggression towards others and self-aggression do not coincide in time; in other words, phases with tendency to use violence towards others or towards oneself, alternate [1].

Several years ago, a discussion arose about whether certain psychotropic drugs, particularly antidepressants of the serotonin reuptake inhibitor (SSRI) type, are associated with an increase in the risk of violence. However, extensive data analyses revealed that it is not possible to distinguish between disease and drug effects [1].

Bipolar Disorders

Symptoms of bipolar disorder (formerly called manic-depressive illness) alternate between manic and depressive phases, often lasting weeks to months, which do not always follow each other directly but may be interrupted by long symptom-free intervals. Some patients are less euphoric and self-exalting during manic phases than irritable, quarrelsome and aggressive.

Bipolar disorders also include some risk profile for aggression and use of violence. The risk is increased by about three times compared to the average population [15]. Increased risk of violence in bipolar patients is also mainly explained by additional alcohol or drug use.

Attention Deficit Hyperactivity Disorder (ADHD)

ADHD is a disorder that has so far been predominantly assigned to childhood and adolescence. However, in many of those affected, this mental disorder can continue into adulthood, albeit often with a modified picture of mental problems. The core symptoms in childhood and adolescence are attention deficits and hyperactivity, disorganized courses of action, uncontrolled impulsivity, and even unpredictable outbursts of rage on minor occasions. Often there is also a disturbance of social behavior, which can persist from childhood into adulthood. About 20% of ADHD children have been diagnosed with an antisocial personality disorder that persists into adulthood [16]; addictive substance use is often an additional complication. However, the vast majority of ADHD patients are not prone to aggression and violence. The incidence of this diagnosis, however, is significantly elevated in prison populations. In juvenile detention centers, the incidence can be assumed to be approximately 45%; the most common offenses are violence and drug-related crimes [17–19].

Brain Injuries and Tissue Defects

These psychosyndromes include damage to brain tissue due to injury, circulatory disturbance or progressive loss (degeneration, atrophy) of brain substance. The clinical symptoms depend on which part of the brain is affected. If parts of the brain are damaged that are responsible for the development and control of aggressive behavior, these being primarily the frontal brain and limbic cortical areas around the amygdala in the middle temporal brain (see Chap. 7, Fig. 7.1), the risk of inadequate emotion control on provocation is increased and the threshold for violence is lowered accordingly. Such damage to the brain structure has been found in a high percentage of violent offenders in prison [20, 21].

Post-traumatic Stress Disorder (PTSD)

Many individuals who have been exposed to massive and upsetting life events subsequently develop post-traumatic stress disorder (PTSD). This often occurs several weeks after the stressful and traumatizing life event and manifests itself predominantly in individuals reacting with emotional breakthroughs in situations where they are reminded of the traumatizing event. Such triggers can be the signal of an ambulance or fire truck, television broadcasts or people reminding them of the event, or even just the physical proximity to the place where the traumatizing event took place. The affected person is then overwhelmed by fear, high-level arousal and, in rare cases, aggressive states of agitation, even violent outbursts. So-called "flashbacks" are also common, in which weak triggering stimuli cause the dramatic event to reappear in the sense of a nightmare.

Extensive studies of this syndrome have been conducted on US Vietnam War veterans [22]. Veterans in whom post-traumatic stress disorder was induced by the events of war experienced greater intensity of anger, had significant problems in regulating feelings of anger, exhibited increased rates of crime and violence, and, most importantly, were also more likely to commit serious violent crimes. The number of combat deployments, the amount of subjectively perceived stress during combat, and the intensity of post-traumatic stress disorder were cited as risk factors for the occurrence of violent crime and weapons offenses in the context of PTSD [22].

Not all individuals who had comparably bad traumatizing experiences developed PTSD. It affected mainly those who already had severe behavioral disorders during childhood and adolescence or who had been abused as children [22]. A 15-fold higher frequency of violent acts within one year was found in war veterans with post-traumatic stress disorder than in veterans without PTSD [23]. The relative frequency of carrying out severe violent acts in traumatized war veterans could also be partly explained by the fact that they owned handguns much more often than other people and were more inclined to use firearms.

Borderline Personality Disorder

People with borderline-type personality disorder suffer from poor control of their emotions and disturbance of interpersonal relationship skills (often they are unable to form strong bonds with a partner), and some of them have increased impulsive aggressiveness. Very often, such individuals had severe traumatizing experiences in early childhood, such as sexual abuse, alcohol-addicted or violent parents, institutional upbringing, or absence of parents due to prolonged illness or imprisonment. The prevalence of clinically expressed borderline symptoms is approximately 1.5% of the general population [24]. The symptoms of these personality disorders differ between men and women; women are more prone to self-injury in the form of superficial skin scratching, which paradoxically provides them with a relaxing effect; men, on the other hand, are more prone to alcohol and drug abuse, and often in conjunction with this, to violent fistfights. It is therefore not surprising that borderline men are more often found in detention centers, and women more often in psychiatric hospitals, to which they are admitted after acts of self-harm. In a study of a large number of murderers in prison, borderline personality disorder was one of the most common psychiatric diagnoses [25].

With the help of functional nuclear magnetic resonance imaging, in borderline patients who had been shown emotional images, the brain center relevant for aggression, namely the amygdala, was found to have become insufficiently regulated and overactive, which was interpreted as the neurobiological basis for an increased impulsive response in these individuals [26, 27].

With newer psychotherapeutic techniques, this personality disorder, which was previously considered difficult to treat, can now be treated much more successfully.

Dissocial/Antisocial Personality Disorders

Dissocial or antisocial personality disorder is diagnosed when socially disruptive behavior is deeply rooted in the personality, begins early in adolescence and is persistent. The disorder occurs in approximately 2% of the population [24]. Dissocial personality traits include irresponsibility, low frustration tolerance, high aggression, self-centeredness, impulsivity and lack of compassion. Such individuals show callous unconcern for the feelings of others, irresponsible attitudes and disregard for social norms, rules and obligations, inability to maintain lasting relationships, a tendency to aggressive and violent behavior in the absence of guilt, and inability to learn from negative experience, especially from punishment. The causes of the development of such antisocial characters lie on the one hand in the early childhood environment, and on the other hand in a genetically determined personality disposition (see Chap. 5).

Psychopathy

For another abnormal character trait related to dissocial personality disorder, the term "psychopathy" was created by the Canadian psychiatrist Hare [28, 29]. Typical of these psychopaths are arrogant-deceptive manners, lack of capacity for self-criticism, lack of empathy and callousness, along with excessive social demands. Such perpetrators have no intelligence deficits, and use other people with emotionless and unrepentant indifference as means to an end. They enforce their interests without any consideration for others, and even with violence, but are less conspicuous for impulsive acts of violence caused by high-grade excitement than for purposeful, goal-oriented actions aimed at exploitation. In doing so, they may well exhibit superficial, winning and tricky behavior.

Many planned, proactive acts of violence are perpetrated by such psychopaths, whose prevalence has been estimated at 1–2% of the male population and who are also said to be found in 4% of corporate managers [30]. Psychopaths account for 20–30% of violent criminals and commit half of all serious crimes [28]. Studies of the brains of such individuals by magnetic resonance imaging revealed structural and functional deficits in the anterior and inferior frontal brain and in the region of the amygdala, thus in the brain areas responsible for the control of emotions and empathy, and for moral behavior [28, 31–33].

Narcissistic and Histrionic Personality Disorders

Narcissistic personality disorder is present in people whose sole purpose in life is to be seen by others as superior, great and unattainable. They want to be constantly surrounded by admiration, and cannot bear it if anyone doubts their greatness, superiority or uniqueness. Occurrence: 0.8% of the population [24]. Most of these perpetrators can certainly perceive the feelings of others, so they do not lack empathy in the strict sense. However, they do not let this influence their decisions and actions, and do not care about the needs of others. The same is true for histrionic (formerly called hysterical) personality disorders.

Because of their pathological need for recognition, narcissists and histrionics are quick to take offense, and then react aggressively and violently. They show considerable deficits in self-esteem regulation as well as misguided strategies for dealing with disappointments and threats to their self-image. In some narcissistically disturbed people, there is a tendency to highly explosive violence in response to actual or perceived slights. This is called "malignant" narcissism.

Paranoid Personality Disorder: Fanatics

Frequency: approx. 1.7% of the population [24]. People with paranoid personality disorder have a hostile attitude towards their fellow human beings, are suspicious, and tend to distort experiences and interpret these as being directed against them. They often have fanatical traits and stubbornly insist on their own opinion to the last. Many have pronounced querulous characteristics, insist uncompromisingly and militantly on actual or assumed own right, without consideration of losses. Similar to narcissists, they have inflated self-esteem and an exaggerated sensitivity to rejection; in addition, they often show pathological jealousy. Many are devotees of conspiracy theories as explanations for particular events, and particular social or political constellations. They include sectarians, power-seeking fighting fanatics and expansive idea fanatics. The German psychiatrist Kretschmer (1888–1964) once said of such people: "In calm times we appraise them, in troubled times they dominate us."

Paranoid personality disorders are not to be confused with paranoid psychoses. In the latter, full-blown delusional symptoms and sensory delusions are present, especially auditory hallucinations, often associated with disorganized thinking and behavior. This degree of disturbance of reality is not found in paranoid personality disorders.

Pathological Irascibility, Rage Syndrome, Cholerics

This psychosyndrome, described in the American classification system of mental disorders (DSM V) as intermittent explosive disorder and listed in the international classification system (ICD 10) under disorders of impulse control as "intermittent irritability", is characterized by intermittent explosive outbursts of rage, which can be triggered by even the most minor events. These people, colloquially referred to as cholerics, become verbally abusive to the point of physical violence, but are largely inconspicuous in the periods between such explosive outbursts. When others are harmed by such outbursts, this type of offender tends to feel remorse and guilt afterwards. The disorder begins in young adulthood. According to a study conducted in the United States, about 4% (12-month prevalence) of the male population is said to suffer from it. This is more than schizophrenic and bipolar individuals combined [34]. Affected individuals often suffer significantly from their disorder and no longer understand their own behavior after violent outbursts. Nevertheless, such outbursts may occur several times a month or even a week. In forensic psychiatry, these individuals are usually classified as affect offenders.

By means of functional magnetic resonance imaging, it has been demonstrated that patients with pathological irascibility have a malfunction of the amygdala and parts of the inferior frontal brain, which controls the activity of the amygdala. If these patients are shown pictures of faces expressing anger, the activity of the amygdala, which responds to danger, is increased and that of the inferior frontal cortex is decreased. The aggression center in the brain of such cholerics can therefore no longer be sufficiently inhibited by the cortical regions responsible for this, even in the case of minor frustrating experiences [35].

Risk of Violence Due to Personality Disorders

If we add up the percentage frequencies of personality disorders that are associated with an increased risk of violence (borderline, narcissistic/histrionic, dissocial, paranoid, pathological irascibility, psychopathy), we arrive at a good 10% of the population. However, since only about 2% of adults become violent offenders (men much more often than women) [36], and since there are not only perpetrators with one of the these personality disorders, but also offenders with other mental disorders such as psychoses, or perpetrators who do not have a psychiatric illness at all and have been psychologically inconspicuous up to now, statistically only a small proportion of people with one of the aforementioned personality disorders become violent offenders during the course of their adult lives. Nevertheless, their risk is said to be ten times higher than that of the average population; it is highest among male borderline patients, especially if increased alcohol or drug consumption is also involved.

At the neurobiological level, personality-disordered violent offenders were found to have reduced activity of the enzyme MAO-A, which is responsible in the brain for the breakdown of drive-enhancing neurotransmitters, and a deficiency of the hormone oxytocin, which is activated during interpersonal affection; in addition, structural and functional deficits in the frontal brain and limbic system were detectable [27, 37].

The above-mentioned types of personality disorders rarely occur in pure form; combinations of two or more forms and coincidence with alcohol or drug addiction are more frequent. The degrees of symptom intensity can vary greatly, and there are smooth transitions to the realm of normal psychology.

References

1. Maier W, Hauth I, Berger M, Saß H (2016) Interpersonal violence in the context of affective and psychotic disorders. Nervenarzt 87(1):53–68. https://doi.org/10.1007/s00115-015-0040-6

2. Elbogen EB, Johnson SC (2009) The intricate link between violence and mental disorder. Arch Gen Psychiatry 66(2):152. https://doi.org/10.1001/archgenpsychiatry.2008.537
3. Sariaslan A, Arseneault L, Larsson H, Lichtenstein P, Fazel S (2020) Risk of subjection to violence and perpetration of violence in persons with psychiatric disorders in Sweden. JAMA Psychiat 77(4):359–367. https://doi.org/10.1001/jamapsychiatry.2019.4275
4. Stevens H, Laursen TM, Mortensen PB, Agerbo E, Dean K (2015) Post-illness-onset risk of offending across the full spectrum of psychiatric disorders. Psychol Med 45(11):2447–2457. https://doi.org/10.1017/S0033291715000458
5. NCI-The National Confidential Inquiry into Suicide and Homicide by People with Mental Illness. Annual Report 2017: England, Northern Ireland, Scotland and Wales. https://www.hqip.org.uk/wp-content/uploads/2018/02/CApw8N.pdf%0D
6. Fazel S, Grann M (2004) Psychiatric morbidity among homicide offenders: a Swedish population study. Am J Psychiatry 161(11):2129–2131. https://doi.org/10.1176/appi.ajp.161.11.2129
7. Chang Z, Larsson H, Lichtenstein P, Fazel S (2015) Psychiatric disorders and violent reoffending: a national cohort study of convicted prisoners in Sweden. Lancet Psychiatry 2(10):891–900. https://doi.org/10.1016/S2215-0366(15)00234-5
8. Wehring HJ, Carpenter WT (2011) Violence and Schizophrenia. Schizophr Bull 37(5):877–878. https://doi.org/10.1093/schbul/sbr094
9. Swanson JW, Swartz MS, Van Dorn RA et al (2006) A national study of violent behavior in persons with schizophrenia. Arch Gen Psychiatry 63(5):490–499. https://doi.org/10.1001/archpsyc.63.5.490
10. Nielssen O, Large M (2010) Rates of homicide during the first episode of psychosis and after treatment: a systematic review and meta-analysis. Schizophr Bull 36(4):702–712. https://doi.org/10.1093/schbul/sbn144
11. Fazel S, Wolf A, Palm C, Lichtenstein P (2014) Violent crime, suicide, and premature mortality in patients with schizophrenia and related disorders: a 38-year total population study in Sweden. Lancet Psychiatry 1(1):44–54. https://doi.org/10.1016/S2215-0366(14)70223-8
12. Mojtabai R (2006) Psychotic-like experiences and interpersonal violence in the general population. Soc Psychiatry Psychiatr Epidemiol 41(3):183–190. https://doi.org/10.1007/s00127-005-0020-4
13. Wilmanns K (1940) Über Morde im Prodromalstadium der Schizophrenie. Z f d ges Neuro u Psych. 170:583–662
14. Fazel S, Wolf A, Chang Z, Larsson H, Goodwin GM, Lichtenstein P (2015) Depression and violence: a Swedish population study. Lancet Psychiatry 2(3):224–232. https://doi.org/10.1016/S2215-0366(14)00128-X
15. Fazel S, Lichtenstein P, Grann M, Goodwin GM, Långström N (2010) Bipolar disorder and violent crime. Arch Gen Psychiatry 67(9):931. https://doi.org/10.1001/archgenpsychiatry.2010.97
16. Mannuzza S, Klein RG, Bessler A, Malloy P, LaPadula M (1993) Adult outcome of hyperactive boys. Educational achievement, occupational rank, and psychiatric status. Arch Gen Psychiatry 50(7):565–576. http://www.ncbi.nlm.nih.gov/pubmed/8317950
17. Mannuzza S, Klein RG, Bessler A, Malloy P, LaPadula M (1998) Adult psychiatric status of hyperactive boys grown up. Am J Psychiatry 155(4):493–498. https://doi.org/10.1176/ajp.155.4.493
18. Witthöft J, Koglin U, Petermann F (2010) Comorbidity of aggressive behavior and ADHD. Kindheit und Entwicklung 19(4):218–227. https://doi.org/10.1026/0942-5403/a000029
19. Sobanski E (2006) Psychiatric comorbidity in adults with attention-deficit/hyperactivity disorder (ADHD). Eur Arch Psychiatry Clin Neurosci 256(S1):i26–i31. https://doi.org/10.1007/s00406-006-1004-4
20. Darby RR, Horn A, Cushman F, Fox MD (2018) Lesion network localization of criminal behavior. Proc Natl Acad Sci 115(3):601–606. https://doi.org/10.1073/pnas.1706587115
21. Schiltz K, Witzel JG, Bausch-Hölterhoff J, Bogerts B (2013) High prevalence of brain pathology in violent prisoners: a qualitative CT and MRI scan study. Eur Arch Psychiatry Clin Neurosci 263(7):607–616. https://doi.org/10.1007/s00406-013-0403-6

References

22. Hiley-Young B, Blake DD, Abueg FR, Rozynko V, Gusman FD (1995) Warzone violence in Vietnam: An examination of premilitary, military, and postmilitary factors in PTSD in-patients. J Trauma Stress 8(1):125–141. https://doi.org/10.1002/jts.2490080109
23. Hahn JW, Aldarondo E, Silverman JG, McCormick MC, Koenen KC (2015) Examining the association between posttraumatic stress disorder and intimate partner violence perpetration. J Fam Violence 30(6):743–752. https://doi.org/10.1007/s10896-015-9710-1
24. Fiedler P (2018) Epidemiology and course of personality disorders. Zeitschrift für Psychiatr Psychol und Psychother. 66(2):85–94. https://doi.org/10.1024/1661-4747/a000344
25. Dixon L, Hamilton-Giachritsis C, Browne K (2008) Classifying partner femicide. J Interpers Violence 23(1):74–93. https://doi.org/10.1177/0886260507307652
26. Herpertz SC, Dietrich TM, Wenning B et al (2001) Evidence of abnormal amygdala functioning in borderline personality disorder: a functional MRI study. Biol Psychiatry 50(4):292–298. http://www.ncbi.nlm.nih.gov/pubmed/11522264
27. Herpertz SC, Nagy K, Ueltzhöffer K et al (2017) Brain mechanisms underlying reactive aggression in borderline personality disorder-sex matters. Biol Psychiatry 82(4):257–266. https://doi.org/10.1016/j.biopsych.2017.02.1175
28. Cornell DG, Warren J, Hawk G, Stafford E, Oram G, Pine D (1996) Psychopathy in instrumental and reactive violent offenders. J Consult Clin Psychol 64(4):783–790. http://www.ncbi.nlm.nih.gov/pubmed/8803369
29. Hare RD, Clark D, Grann M, Thornton D (2000) Psychopathy and the predictive validity of the PCL-R: an international perspective. Behav Sci Law 18(5):623–645. http://www.ncbi.nlm.nih.gov/pubmed/11113965
30. Babiak P, Neumann CS, Hare RD Corporate psychopathy: talking the walk. Behav Sci Law 28(2):174–193. https://doi.org/10.1002/bsl.925
31. de Oliveira-Souza R, Hare RD, Bramati IE et al (2008) Psychopathy as a disorder of the moral brain: fronto-temporo-limbic grey matter reductions demonstrated by voxel-based morphometry. Neuroimage 40(3):1202–1213. https://doi.org/10.1016/j.neuroimage.2007.12.054
32. Decety J, Chen C, Harenski CL, Kiehl KA (2015) Socioemotional processing of morally-laden behavior and their consequences on others in forensic psychopaths. Hum Brain Mapp 36(6):2015–2026. https://doi.org/10.1002/hbm.22752
33. Walters GD, Kiehl KA (2015) Limbic correlates of fearlessness and disinhibition in incarcerated youth: exploring the brain-behavior relationship with the hare psychopathy checklist: youth version. Psychiatry Res 230(2):205–210. https://doi.org/10.1016/j.psychres.2015.08.041
34. Kessler RC, Coccaro EF, Fava M, Jaeger S, Jin R, Walters E (2006) The prevalence and correlates of DSM-IV intermittent explosive disorder in the national comorbidity survey replication. Arch Gen Psychiatry 63(6):669. https://doi.org/10.1001/archpsyc.63.6.669
35. Coccaro EF, Fitzgerald DA, Lee R, McCloskey M, Phan KL (2016) Frontolimbic morphometric abnormalities in intermittent explosive disorder and aggression. Biol Psychiatry Cogn Neurosci Neuroimaging 1(1):32–38. https://doi.org/10.1016/j.bpsc.2015.09.006
36. Boers K (2019) Delinquenz im Altersverlauf. Monatsschrift für Kriminologie und Strafrechtsreform 102(1):3–42. https://doi.org/10.1515/mks-2019-0004
37. Mancke F, Herpertz SC, Bertsch K (2018) Correlates of aggression in personality disorders: an update. Curr Psychiatry Rep 20(8):53. https://doi.org/10.1007/s11920-018-0929-4

Chapter 11
Alcohol, Drugs and Violence

Without alcohol and drugs, the problem of violence would not be eradicated, but it would be less present. What are the relationships between the abuse of addictive substances and violence?

Addiction as Cause and Consequence of Violence

Genetic factors and early childhood experiences play just as important a role in the development of addiction as they do in the predisposition to violence. The experience of early violence or neglect in childhood often results in substance abuse in those affected, which in turn predisposes them to behave violently later in life. One-third to half of all people treated for alcohol or drug dependency have experienced violence or neglect in childhood [1]. Addiction can occur as a result of violence, with alcohol or drugs being consumed as a form of self-medication to repress psychological traumatization. In families where parents are alcohol or drug addicted, violence against children is particularly common. This in turn increases the risk of the offspring becoming addicted and violent themselves later on. Addiction can thus be both a cause and a consequence of violence [1].

Among the mental disorders that can lead to acts of violence, alcohol and drug addiction play the most important role. In the Swedish study mentioned in Chap. 10 [2], the greatest risk of violence was found for addicts: over the course of one year, 25 out of 1000 drug addicts and 13 out of 1000 alcoholics became violent. It should be noted that alcohol or drug addiction in the majority of those affected develops on the basis of another psychiatric illness or personality disorder, and usually the combination with substance abuse is associated with an increased risk of violence. The occurrence of violence in mental illness without additional alcohol or drug consumption is significantly lower [2].

Frequency of Violence Under the Influence of Alcohol

Alcohol is regarded as an essential enrichment of celebrations and social events, loosens the mood for many, relaxes and makes communication more unconstrained. However, its disinhibiting, destructive, antisocial, aggression-promoting and violence-facilitating properties are the other side of the coin [3].

For Europe and North America, up to 50–60% of homicides have been reported to be committed under the influence of alcohol, and for Russia, over 70% [3, 4]. In Germany, according to police crime statistics for 2018, more than a quarter of violent crime in the form of manslaughter, murder, assault and rape was committed under the influence of alcohol. This is particularly true of grievous and actual bodily harm [5]. Vandalism and brawls are regularly accompanied by alcohol consumption. There is a worldwide correlation between alcohol consumption per capita and the frequency of violence; the more people drink, the more often they are injured by violent acts [6, 7].

According to the WHO, Europe is the world region with the highest per-capita alcohol consumption. More than twice as much is consumed here as elsewhere [8]. The annual cost of all violent crimes involving alcohol was estimated at 33 billion euros for the EU in 2003 [8, 9].

In Europe, four out of ten murders and one in every six suicides are alcohol related [9]. However, depending on the type of statistical survey and the country studied, data on rates of alcohol-related violence vary considerably. For example, for northern European countries such as Sweden and Norway, 80–90% of all violent crimes were reported to occur under the influence of alcohol, in France and Germany about a quarter, and in Spain just under half [9]. Alcohol also plays an inglorious role in domestic and sexual violence. In 2002, 29% of all rape cases were related to alcohol consumption [9]. Half of all domestic violence cases recorded in 2014 happened under the influence of alcohol, with a high number of unreported cases. Even higher figures are reported for northern European countries such as Ireland and Iceland [9, 10]. In the UK, nearly 60% of convicted rapists in prison are reported to have consumed alcohol before committing the crime [10]. Violence against children is also often alcohol induced. Thirty-two percent of all fatal child abuse within Germany occurred under the influence of alcohol [10].

By no means all people who have alcohol problems or are addicted to alcohol tend to commit violent acts. A genetically and/or biographically derivable disposition to violence is present in those individuals who become aggressive under the influence of alcohol [5]. In these individuals, alcohol increases aggressiveness and reduces inhibition mechanisms; alternatives for action are no longer apparent to violent offenders [3, 5]. The more alcohol a person consumes, the higher the risk that they will become a perpetrator or also a victim of violence. Domestic violence and relationship violence in particular play out on the ground of an addiction problem, with victim and perpetrator status often changing and mutual complaining behavior being typical.

Effect of Alcohol on the Brain

Alcohol has several mechanisms of attack in the brain. It interferes with the activity of both the inhibitory neuronal transmitter GABA (gamma-aminobutyric acid) and the stimulatory neurotransmitter glutamate, upsetting the balance between inhibitory and activating neuronal functions. This is especially true for the limbic system, which is responsible for the neuronal modulation of emotional responses, as well as for the frontal brain, which controls the activities of the limbic system [5, 11, 12].

Alcohol, like all addictive substances, also activates the brain's reward system with the help of the neurotransmitter dopamine. This explains the frequently encountered elevated mood under moderate alcohol consumption. On the other hand, in vulnerable individuals in whom the neurotransmitter serotonin is decreased, alcohol impairs the inhibitory effect of the frontal brain on elementary emotional functions of the limbic system, which favors the manifestation of aggressive behavior [5, 11]. By means of functional magnetic resonance imaging studies of the brain, it has been shown that alcohol, when a person is provoked, increases the activity of the amygdala, which explains the increased aggressiveness. By stimulating the activity of the brain's reward center, the nucleus accumbens, alcohol not only elevates mood, but also increases drive and aggressiveness in predisposed individuals [13].

Inhaled substances have a similar effect to alcohol; like the latter, they can cause significant damage to brain tissue with long-term use.

Effects of Drugs

Drugs are addictive because they act directly on the brain's reward system, thereby creating a satisfied-uplifted mood and prompting the brain to repeat the action that led to it, hence further drug use. Addicts cannot resist the urge of repeated drug use even with knowledge of the harmful long-term consequences. Drugs also have disinhibiting and aggression-promoting properties similar to alcohol. However, these vary depending on the type of drug and whether the user is currently under the influence of drugs, is in a withdrawal phase or whether personality changes or even damage to brain tissue has already occurred as a result of long-term drug use.

Cannabis has so far been credited with a predominantly calming and neutral effect. However, recent reviews evaluating all publications on the subject of violence and cannabis have come to the conclusion that cannabis consumption is also associated with an increased risk of violence [14]. Many users show increased aggressiveness within the first two weeks of cannabis withdrawal [15]. In addition, users of cannabis run the risk of developing schizophrenia-like psychosis if they continue to use it, which in turn increases the risk of aggressive behavior.

Hallucinogens (e.g., LSD, psilocybin) have a stimulating effect on the serotonin receptors in the brain, which explains the disturbing effects of these substances on mood. Hallucinogens have even been reported to have an aggression- and anxiety-reducing effect [15, 16]. However, as the name suggests, they evoke disturbances of reality ranging from visual distortions of perception up to hallucinations.

Opiates (e.g., heroin) have a calming and euphoric effect immediately after ingestion. However, dependency develops very quickly, often associated with unbearable psychological and physical withdrawal symptoms and resulting drug-related crime, including acquisition crime. The increased rate of violence among heroin addicts is partly explained by this, and partly by the fact that many also have an antisocial personality disorder even before the start of their drug career [15]. Opiate addiction is also associated with a high risk of suicide.

Psychostimulants: The risk of violence from drugs is particularly high in the case of psychostimulant substances such as cocaine. Crystal meth (= methamphetamine), cocaine and crack, which is made from cocaine, salt and baking soda, and results in a particularly rapid high level of dependence, belong to this group of drugs. In low doses and in the initial phase of use, they help to create a euphoric mood with overconfidence and increased drive. In the advanced phase, a dependency develops that requires higher and higher doses to achieve this effect. Frequent consequences range from psychoses with persecution and threat delusions and aggressive states of excitement with breakthroughs that are dangerous to oneself and others, up to and including serious violent offenses. Crystal-meth intoxication can lead to sudden violent breakthroughs with subsequent memory lapse, associated with depression, anxiety reactions and hallucinations [17].

The close connection between alcohol or drug addiction and violent acts was confirmed in a very extensive summary of 35 meta-analyses on this topic [4]. High-risk groups among addicts are young men and persons who additionally suffer from psychotic disorders.

Drug Terror

Violence is often practiced by drug dealers, not only to collect debts, but also as a warning against turning away from the drug scene and reporting drug trafficking to the authorities.

Drug wars with ruthless terror occur regularly between gangs and drug cartels in the context of territorial battles that the cartels and their paramilitary units fight among themselves or with the police and military. This situation is particularly dramatic in some Central and South American countries, which thus have the highest levels of violence in the world [18]. In Mexico, for example, one-third to half of all murders bear the hallmarks of drug cartels and related organized crime [19]. The bloodiest years there so far have been 2018, with 23,000 deaths, and 2019, with 34,000 deaths [20]. In terms of population size, this is equivalent to over 30 times more deaths from drug terror alone than the total number of all homicides in Western Europe. Add to that over 73,000 missing persons in Mexico, and up to 50,000 children orphaned there in 2020 as a result of drug wars [20]. Similar conditions exist in some other Central and South American countries [18].

References

1. Schäfer I, Pawils S, Driessen M et al (2017) Understanding the role of childhood abuse and neglect as a cause and consequence of substance abuse: the German CANSAS network. Eur J Psychotraumatol 8(1):1304114. https://doi.org/10.1080/20008198.2017.1304114
2. Sariaslan A, Arseneault L, Larsson H, Lichtenstein P, Fazel S (2020) Risk of subjection to violence and perpetration of violence in persons with psychiatric disorders in Sweden. JAMA Psychiat 77(4):359–367. https://doi.org/10.1001/jamapsychiatry.2019.4275
3. Parrott DJ, Eckhardt CI (2018) Effects of alcohol on human aggression. Curr Opin Psychol 19:1–5. https://doi.org/10.1016/j.copsyc.2017.03.023
4. Duke AA, Smith KMZ, Oberleitner LMS, Westphal A, McKee SA (2018) Alcohol, drugs, and violence: a meta-meta-analysis. Psychol Violence 8(2):238–249. https://doi.org/10.1037/vio0000106
5. Beck A, Heinz A (2013) Alcohol-related aggression. Dtsch Aerzteblatt Online. 110(42):711–715. https://doi.org/10.3238/arztebl.2013.0711
6. Bye EK, Rossow I (2009) The impact of drinking pattern on alcohol-related violence among adolescents: an international comparative analysis. Drug Alcohol Rev 29(2):131–137. https://doi.org/10.1111/j.1465-3362.2009.00117.x
7. Cherpitel CJ, Witbrodt J, Ye Y, Korcha R (2018) A multi-level analysis of emergency department data on drinking patterns, alcohol policy and cause of injury in 28 countries. Drug Alcohol Depend 192:172–178. https://doi.org/10.1016/j.drugalcdep.2018.07.033
8. Graham L, Parkes T, McAuley A, Doi L (2012) Alcohol problems in the criminal justice system: an opportunity for intervention. World Health Organization. https://www.euro.who.int/__data/assets/pdf_file/0006/181068/e96751-ver-2.pdf. Published 2012. Accessed 13 Jan 2021
9. Anderson P, Baumberg B (2006) Alcohol in Europe, A public health perspective; A report for the European Commission. London. https://ec.europa.eu/health/archive/ph_determinants/life_style/alcohol/documents/alcohol_europe_en.pdf
10. World Health Organization RO for E (2005) Alcohol and interpersonal violence: policy briefing. https://apps.who.int/iris/handle/10665/107351. Published 2005. Accessed 18 Jan 2021
11. Miczek KA, DeBold JF, Hwa LS, Newman EL, de Almeida RMM (2015) Alcohol and violence: neuropeptidergic modulation of monoamine systems. Ann N Y Acad Sci 1349:96–118. https://doi.org/10.1111/nyas.12862
12. Heinz AJ, Beck A, Meyer-Lindenberg A, Sterzer P, Heinz A (2011) Cognitive and neurobiological mechanisms of alcohol-related aggression. Nat Rev Neurosci 12(7):400–413. https://doi.org/10.1038/nrn3042
13. Gan G, Sterzer P, Marxen M, Zimmermann US, Smolka MN (2015) Neural and behavioral correlates of alcohol-induced aggression under provocation. Neuropsychopharmacology 40(13):2886–2896. https://doi.org/10.1038/npp.2015.141
14. Dellazizzo L, Potvin S, Athanassiou M, Dumais A (2020) Violence and cannabis use: a focused review of a forgotten aspect in the era of liberalizing cannabis. Front Psychiatry 11. https://doi.org/10.3389/fpsyt.2020.567887
15. Tomlinson MF, Brown M, Hoaken PNS (2016) Recreational drug use and human aggressive behavior: A comprehensive review since 2003. Aggress Violent Behav 27(April 2019):9–29. https://doi.org/10.1016/j.avb.2016.02.004
16. Walsh Z, Hendricks PS, Smith S et al (2016) Hallucinogen use and intimate partner violence: prospective evidence consistent with protective effects among men with histories of problematic substance use. J Psychopharmacol 30(7):601–607. https://doi.org/10.1177/0269881116642538

17. Payer DE, Lieberman MD, London ED (2011) Neural correlates of affect processing and aggression in methamphetamine dependence. Arch Gen Psychiatry 68(3):271. https://doi.org/10.1001/archgenpsychiatry.2010.154
18. Rosen JD, Kassab HS (2019) Drugs, gangs, and violence. Palgrave Macmillan
19. Heinle K, Molzahn C, Shirk D (2015) Drug violence in Mexico. https://justiceinmexico.org/wp-content/uploads/2015/04/2015-Drug-Violence-in-Mexico-final.pdf. Published 2015. Accessed 12 May 2017
20. Wikipedia.org. Mexican drug war. https://en.wikipedia.org/wiki/Mexican_drug_war. Accessed 26 Apr 2021

Chapter 12
Psychology of Violence

Compared to the brain-biological approaches to explaining the occurrence of violence, which has only become possible in recent decades with the rapid development of neuroscience, especially brain imaging techniques, psychological theories of violence have a long tradition. They play a key role in understanding the phenomenon.

Historical Attempts to Explain Violence

The Greek historian Thucydides (460–395 BCE) has already provided us with some thoughts on the causes of interpersonal violence. He was also a military strategist in the Peloponnesian War (431–404 BCE), in which Athens, Sparta and Thebes fought for supremacy and which ultimately led to the downfall of classical Greece. Thucydides (Fig. 12.1), with his excellent understanding of the situation at that time, analyzed the course and background of this war and came to the conclusion that man by nature strives for power and glory and spares no crime to achieve this goal [1].

The same view was expressed about a thousand years later by the English constitutional lawyer, mathematician and philosopher Thomas Hobbes (1588–1679) in his work *Leviathan*, published in 1651 [2]. He identified three causes of man's tendency to violence: competition, insecurity and addiction to fame. Hobbes (Fig. 12.2) was affected by the atrocities of the English Civil War (1642–1651) and believed that a state of anarchy and the accompanying violence could only be contained by a strong state authority holding the monopoly on violence. By severely punishing violent acts, this authority ("Leviathan") would have the effect of deterring violence. Hobbes is credited with the oft-quoted phrase *homo homini lupus* (man is wolf to man) [2].

In contrast to Thucydides and Hobbes, the Genevan philosopher and educator Jean-Jaques Rousseau (1712–1778) (Fig. 12.3) believed that man in his natural form was inherently good and peaceable. The concept of the "noble savage" goes back to Rousseau. It was only later processes of civilization and the increasingly competitive thinking that went along with them that gave rise to resentment, hatred and aggression

Fig. 12.1 Thucydides (460–396 BCE). Thucydides; User: Shakko, created on 01/01/2008; Available online at: https://de.wikisource.org/wiki/Thukydides#/media/Datei:Thucydides_pushkin01.jpg; (retrieved October 2020)

Fig. 12.2 Thomas Hobbes (1588–1679). Thomas Hobbes (detail from a painting by John Micheal Wright, circa 1669–1670); Public domain; Available online at: https://de.wikipedia.org/wiki/Thomas_Hobbes#/media/Datei:Thomas_Hobbes_by_John_Michael_Wright_(2).jpg; (retrieved October 2020)

and thus created the conditions for violent conflicts. Culture and civilization had alienated man from his original good-hearted nature [3].

Rousseau thus also contradicted the Church's doctrine, according to which man is predisposed to evil through original sin, which becomes active in the form of the devil.

The Italian forensic pathologist and psychiatrist Cesare Lombroso (1853–1909) believed, based on his studies of criminals, that their tendency to criminal behavior could be recognized not only by their psychological but also by their physical constitution. He founded the doctrine of the "born criminal," whose predisposition to criminal behavior was anchored in his personality, because it was inherited. According to Lombroso (Fig. 12.4), this type of person was characterized by a special skull

Fig. 12.3 Jean-Jaques Rousseau (1712–1778). Jean-Jaques Rousseau, pastel by Maurice Quentin de La Tour, 1753; Public domain; Available online at: https://de.wikipedia.org/wiki/Jean-Jacques_Rousseau#/media/Datei:Jean-Jacques_Rousseau_(painted_portrait).jpg; (retrieved Oct. 2020)

Fig. 12.4 Cesare Lombroso (1835–1909). Cesare Lombroso, User: MoritzB., created on 17/09/2007; Available online at: https://de.wikipedia.org/wiki/Cesare_Lombroso#/media/Datei: Lombroso.JPG; (retrieved Oct. 2020)

shape such as a low forehead and fused eyebrows. Such physical features were signs of an inferior and more violent stage of human development and of a deeply rooted predisposition to criminality [4]. Lombroso's theories could be partially confirmed by later scientific investigations insofar as heredity (thus an a priori existing personality disposition) can explain a not insignificant part of the multiple factors contributing to violent character (see Chap. 5). His view that the physical constitution, in particular the shape of the head, is relevant for this, however, sounds bizarre today; in any case, it did not stand up to critical scientific investigations [4].

The Drive Theories of Freud and Lorenz

Does violence have a drive character or is it an understandable reaction of a human being who is by nature more inclined to peacefulness and who has to defend himself against external threats?

The drive theory was advanced by the psychoanalyst Sigmund Freud (1865–1939) in his writings *Jenseits des Lustprinzips* (*Beyond the Pleasure Principle*, 1920) [5], *Das Unbehagen in der Kultur* (*The Discomfort in Culture*, 1930) [6] and *Warum Krieg* (*Why War*, 1933) [7]. Freud, who emigrated from Nazi Germany to London (several of his close relatives fell victim to the Holocaust) did not address the issue of aggression and violence until his late writings. Prior to that, his life's work centered on the development of psychoanalysis, with sexuality, especially repressed infantile sexual stages, playing a central role as the cause of later psychological conflicts. Only in the late work did Freud suspect a death drive (Thanatos), which he saw as antagonistic to the life drive (Eros). The fate of mankind depended on the mastery of the instinct of aggression and destruction he postulated.

A second representative of the drive theory of aggression is Konrad Lorenz (1903–1989). Lorenz assumed that aggressiveness is a spontaneously developing drive which accumulates more and more and therefore has to be vented in a form as harmless as possible, comparable to a steam boiler which overheats and explodes if the pressure is not released. In his book *Das sogenannte Böse—Zur Naturgeschichte der Aggression* (*The So-called Evil—On the Natural History of Aggression*) [8, 9], Lorenz attributed a considerable phylogenetic selection value to aggression and attacks against one's own species. Lorenz was an animal behaviorist and regarded essential principles of animal-aggressive behavior as also transferable to humans. A plausible reason for the phylogenetic survival advantage of violence against one's own species was that after establishing a hierarchy, the successfully more aggressive (male) conspecific secured the right to reproduce for himself and thus passed on both his physically superior and more aggressive nature to his offspring with his genes. The same was true for collective group violence against conspecifics. The psychologically superior and more aggressive group and its subsequent generations with the same mentality ultimately had a higher reproductive fitness and prevailed in phylogenesis [9].

The drive models of Freud and Lorenz were criticized because in addition to a genetic disposition, a desire for violence does not accumulate over time like hunger and thirst, but rather depends on many biographical and situational circumstances that inhibit or promote aggressive behavior.

Frustration Theory and Learning Theory

The aggression-frustration theory, which was developed by the American psychologists and behavior theorists Miller and Dollard [10], states that aggression mainly arises as a result of frustration. If a person is hindered (frustrated) on the way to achieving a desired goal by another, they react with aggression to get the interferer out of the way. One of the consequences of this influential theory was that many educators took the view that children should be brought up with as little frustration as possible in order to turn them into peaceful adults, which led to styles of education lacking boundaries. Not infrequently, such children later became particularly intolerable and aggressive contemporaries.

Another behavioral attempt to explain the development of aggression was presented by the American psychologist Albert Bandura [11]. According to him, aggressive role models in the family or among friends are imitated by the child or adolescent. For example, when a son sees his aggressive father successfully asserting himself against family members or others, he tries this himself. If this behavior is successful, as with the father, it is self-reinforced and maintained. Violent "heroes" in the media can serve similar role model functions.

Violence: A Product of Civilization?

The German-American psychoanalyst Erich Fromm (1900–1980) opposed the view of Freud and Lorenz that there is an aggression instinct that builds up spontaneously and seeks its victims. Fromm, who emigrated from Nazi Germany to the US as a young man, proposed in his book *The Anatomy of Human Destructiveness* [12] that violence is not the result of innate drive-like behavior. Like Rousseau before him, Fromm was of the opinion that wars only occur in more developed societies as a result of realistic conflicts of interest. In contrast to other living beings, only man could become the destroyer of his own kind. He assumed that an a priori existing tendency to violence in man, which would inevitably lead to warlike conflicts, did not exist.

Fromm's thesis of the unique position of humans with regard to collective violence against their own species could be refuted by the primatologist Goodall [13]. She impressively described how chimpanzees wage "war" against neighboring groups with deadly confrontations. Similar observations were made with other ape species and also with predators living in packs like wolves, hyenas and lions, and even with rat tribes (see Chap. 4).

The Banality of Evil

The thesis of the banality of evil occupied large parts of psychology and public opinion after the Second World War and the Nazi period. It was developed by Hannah Arendt on the occasion of the 1961 trial in Jerusalem of Adolf Eichmann, the organizer of the Holocaust [14]. Eichmann presented himself during the trial as a completely normal civil servant who only fulfilled his assigned tasks and thus—without feeling personal enmity towards the victims—organized the extermination process of six million Jews.

Eichmann had many helpers and assistants, and one wonders how it was possible that in civilian life inconspicuous, completely normal men in the areas conquered by the Wehrmacht, especially behind the Eastern Front and in concentration camps, were capable of committing violent crimes of hitherto unknown dimensions. The most common explanations for this were that such mass murders by men who were inconspicuous at home were only possible due to their willingness to obey and the accompanying performance of orders combined with feelings of power and superiority, the shifting of responsibility to superiors, and ways of acting that conformed to the group.

These explanations were supported by two much cited psychological experiments: the Milgram experiment [15], [16] and the Stanford prison experiment [17, 18].

In the former, 40 psychologically normal test subjects were asked to inflict pain on others by electric shocks—in an experiment that portrayed this as a scientific necessity. The subjects were told that the electric shocks were of increasing strength. However, the electric shocks were faked and the persons receiving the shocks were actors who simulated the pain as the intensity of the current increased by screaming with increasing volume. However, the experimental subjects believed the electric shocks and pain were real. Twenty-six of them followed the instructions of the experimenter and went up to the maximum claimed current of 450 V despite intense screams from the actor and his pleas to stop the experiment; only 14 of the 40 subjects stopped before that.

The conclusion was that two-thirds of the test subjects were able to switch off their own conscience due to the instructions of the supervisor (in this case the experimenter) and thus carried out the instructions particularly brutally up to the highest level of pain. A state of moral freedom developed in which one's own responsibility for one's actions was handed over to authority figures and only orders were obeyed.

The Stanford prison experiment was a prison simulation study in which randomly selected student volunteers were divided into two groups. One was to assume the role of a prisoner in a jail, and the other the role of a prison guard. Both groups were taken to a simulated prison where they were required to perform their roles. Former laboratory rooms in the basement of Stanford University were used as prison cells, with doors replaced by specially made lattice doors. After a short time, the "prison guards" behaved so violently and brutally towards the group of "prisoners," who were also selected at random, that the whole experiment had to be stopped after six days [17, 18].

The experiment showed that the granting of power and controlling authority can drastically lower the threshold for devaluing, oppressing and mistreating others. The assigned role of the "jailers" and conformity behaviors within the jailer group facilitated the harassment of subordinates, up to and including sadistic behaviors. The result of the experiment suggests that there is a tendency in every normal citizen to bully others out of a feeling of power and arrogance and that this tendency is also acted out by many if the opportunity for this arises within a suitable system and remains unpunished.

The Milgram experiment and the Stanford prison experiment were criticized because their methods and protocols appeared questionable [16]. Milgram's findings, however, were essentially confirmed by a later follow-up [19]. An experimental set-up replicating the Stanford prison experiment followed a different course. The prisoners organized a resistance to the guards' mistreatment; three prisoners and one guard set up a draconian rule system, which is why this study also was terminated.

Both experiments have been and continue to be confirmed by reality many times. Humiliation and torture of the defeated, of prisoners, deviants and dissenters are among the recurring dark pages of world history. From ancient times to the Inquisition of the Middle Ages to the camps of National Socialism and Stalinism, humiliation and torture can be found not only in the dungeons of dictatorial regimes, but also in special prisons of some democratic states. The best-known more recent examples of this are the prisons operated by the US military, Abu Ghraib [20] and Guantanamo [21].

The disturbing thesis derived from the experiments described—that a majority of people are potential perpetrators of violence and that mistreatment, torture and murder can be attributed to the power of the respective social circumstances or to a mechanical obedience or group pressure—does not relativize the responsibility of the perpetrators. Most accepted the violence-justifying ideology of the commanding system, knew exactly what they were doing, were convinced they were doing the right thing in feeling they were the master race and represented this offensively. Although these experiments have provided fundamental evidence of the importance of the social situation for violent action, critical analyses of the Milgram and Stanford prison experiments [22, 23], as well as recent work on National Socialism, argue [24] that the seemingly superficial banality of evil is instead based on a process of normalization of the use of violence that is the product of a complex interactive dynamic between the person performing it and the social environment [25].

New Psychological Theories of Aggression

The classical theories of aggression have been gradually replaced by newer models taking into account the fact that violence cannot be viewed in a one-dimensional way, as is the case, for example, with drive theory or frustration theory, but is the result of a complex set of conditions in which many factors are involved. An influential recent theory is the general aggression model (GAM) [26, 27], which assumes that genetic and brain-biological preconditions (without further differentiating them) as well as the social, biographical, societal and political background as "distal,", that is, more background, causes shape a personality's disposition to violence. To these "distal" factors must then be added so-called "proximal,", or nearby factors. The latter include the current situation triggering aggression, such as provocation, exclusion, stress or even pain. If distal and proximal cause bundles come together, then initially innerpsychic cognitive and emotional evaluation processes of the situation are activated by the affected person. Depending on the subjective evaluation, the current state of mind, the level of arousal and the person's psychological and physical resources, as well as the assessment of the possible outcome of an aggressive act, it is then carried out or not. If the personality resources are not sufficient to adequately cope with an aggression-triggering situation, an impulsive act (affect act) may occur; if personality resources are sufficient, an impulsive-reactive act of violence may be suppressed and a more thoughtfully planned aggressive act may follow. A simplified schematic representation of the general aggression model is given in Fig. 12.5.

The general aggression model (GAM) has been extended by the so-called I^3 (I-cubed) model of aggression [28]. This model assumes that initial conditions from three different levels are required for the occurrence of an aggressive act. The first level is the personality disposition ("impellance"), which is associated with a different threshold for the propensity to violence; the second level is the intensity of environmental factors that trigger aggression ("instigation"). Here we find an analogy with the disposition-stress model for mental illnesses, according to which a mental state of emergency is always the result of a pre-existing disposition (vulnerability) and an acute or chronic stress situation. In the I^3 model, either intrapsychic or environment-related violence-inhibiting factors ("inhibition") are added as a third component. Inhibiting circumstances can be intrapsychic, such as empathy or fear of consequences, or external circumstances that discourage the perpetration of violence. The

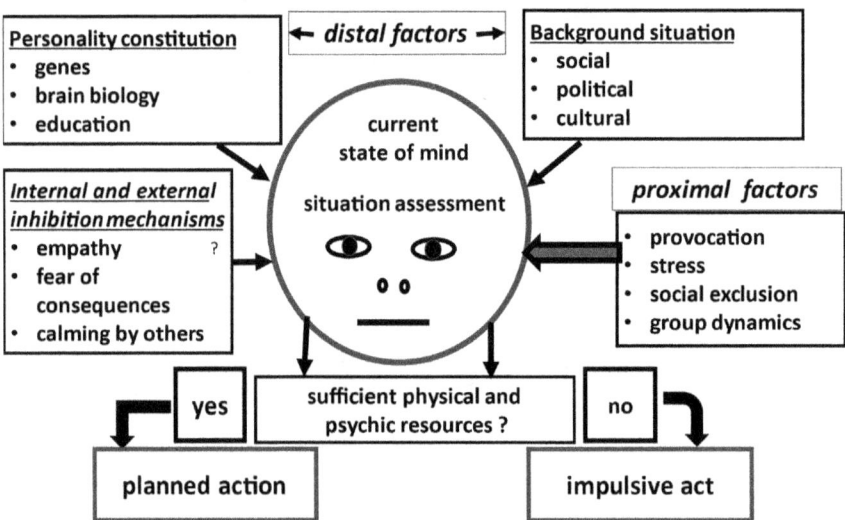

Fig. 12.5 Simplified representation of the general aggression model and the I^3 model of aggression developed from it [28]. Distal factors, proximal factors, type and intensity of the situation triggering aggression, inhibition mechanisms, and the available mental and physical resources determine whether violence is used and which form of violence takes place. Own illustration

occurrence of an act of aggression is thus determined three-dimensionally: by the pre-existing personality disposition, the intensity of aggression-triggering situations, and aggression-inhibiting inner-psychic or external influences.

The Dark Tetrad of Personality

A tetrad is the term used to describe a four-part structure. In the search for personality traits that are particularly conspicuous for dissocial, reckless, aggressive and violent behavior, four defined character traits have been grouped together as the dark tetrad of personality [29]. Not only a great deal of white-collar crime and fraud, but also the use of violence in all its varieties can be considered the work of this dark tetrad. These four components are:

- narcissism
- Machiavellianism
- psychopathy
- sadism.

Narcissists place the admiration of themselves above all other things, place pathological value on being admired as the biggest, best, most beautiful, etc. and subordinate everything else to this endeavor without regard to the needs of others. Narcissists see themselves as the center of the world and react in an abusively aggressive manner when they feel their fragile self-esteem is hurt (see Chap. 10).

Machiavelli (died 1527) was an Italian politician and writer whose name is today associated with ruthless power politics. Machiavellianism is an attitude in which everything that leads to power and serves to maintain it is justified. The end justifies the means. The resulting suffering of others is irrelevant.

Psychopaths know no other values, are characterized by a lack of empathy, by coldness of feeling and, like Machiavellians, by ruthless assertion of their own interests. However, the focus is less on the desire for power than on personal advantage in general. Typical are arrogant-deceptive manners, which such persons know very well how to use to manipulate others. They are less conspicuous for impulsive acts of violence than for unscrupulous, purposeful, goal-oriented actions aimed at exploitation (see Chap. 10).

Sadists enjoy torturing, commit violence for their own sake and can be enthusiastic about orgies of violence. Suffering and pain of others is connected with pleasure for them, which they serve when a suitable opportunity presents itself and punishment is not to be feared (see Chap. 13).

How often are such sinister characters to be found in the population? Pathological narcissists are said to make up about 1% of the population, psychopaths about 2% [30]. Machiavellians are likely to be much more common, especially among men, sadists rather less so. However, an increased risk of violence is also present in several psychosyndromes that are not included in the dark tetrad. These include cholerics and fanatics, paranoids, borderline individuals and psychotics (see Chap. 10).

A bit of dark tetrad is carried by everyone, even if it is only trace elements. Pretty much everyone thinks of themselves as something special, which is an element of narcissism that would be bad if a healthy self-esteem had not been imparted to them in their upbringing. Who has not at some time put their own interests above those of others without taking into account their concerns, and thus been a little Machiavellian? Many may also have experienced a tendency to manipulation of others for their own advantage (i.e., trace elements of psychopathy). And *Schadenfreude*, which everybody knows from their own experience, is the little sister of sadism. From the normal forms of such generally encountered human traits there is a smooth, continuous transition to the extreme variants in the form of personality disorders to which pathological value can be attributed. These are characterized by the fact that either society or the affected persons themselves suffer from them.

There is also a bright tetrad of personality: empathy, selflessness, compassion, affection. At present, the bright tetrad is much more visible in our lives than the dark one. This has not always been the case and is not guaranteed for the whole of our future.

References

1. Wikipedia.org. Thucydides. https://en.wikipedia.org/wiki/Thucydides. Accessed 28 Apr 2021
2. Wikipedia.org. Thomas Hobbes. https://en.wikipedia.org/wiki/Thomas_Hobbes. Accessed 28 Apr 2021
3. Wikipedia.org. Jean-Jacques Rousseau. https://en.wikipedia.org/wiki/Jean-Jacques_Rousseau. Accessed 26 Apr 2021
4. Wikipedia.org. Cesare Lombroso. https://en.wikipedia.org/wiki/Cesare_Lombroso. Accessed 26 Apr 2021
5. Freud S (1920) Jenseits Des Lustprinzips. Fischer Verlag, Frankfurt am Main
6. Freud S (1930) Das Unbehagen in Der Kultur. Internationaler Psychoanalytischer Verlag, Wien
7. Freud S (1933) Warum Krieg? Briefwechsel Mit Albert Einstein. Diogenes
8. Lorenz K (1963) Das Sogenannte Böse: Zur Naturgeschichte Der Aggression. Dr. G. Borotha-Schoeler Verlag, Wien
9. Wikipedia.org, Lorenz K, On aggression. https://en.wikipedia.org/wiki/On_Aggression. Accessed 26 Apr 2021
10. Dollard J, Miller NE, Doob LW, Mowrer OH, Sears RR (1939) Frustration and aggression. Yale University Press, New Haven. https://doi.org/10.1037/10022-000
11. Bandura, Albert; Walters RH (1959) Adolescent aggression: a study of the influence of child-training practices and family interrelationships. Ronald Press New York
12. Fromm E (1973) The anatomy of human destructiveness. Holt, Rinehart and Winston, New York
13. Goodall J (1971) In the shadow of man. William Collins Sons & Co., London
14. Arendt H (1963) Eichmann in Jerusalem: a report on the banality of evil. Penguin Press, New York
15. Milgram S (1974) Obedience to authority. An experimental view. Harper & Row, New York
16. Perry G (2013) Behind the shock machine: the untold story of the notorious Milgram psychology experiments, Revised. The New Press, New York
17. Wikipedia.org. Stanford prison experiment. https://en.wikipedia.org/wiki/Stanford_prison_experiment. Accessed 26 Apr 2021

References

18. Haney C, Banks W, Zimbardo PG (1973) A study of prisoners and guards in a simulated prison. Nav Res Rev 30:4–17
19. Burger JM (2009) Replicating Milgram: would people still obey today? Am Psychol 64(1):1–11. https://doi.org/10.1037/a0010932
20. Wikipedia.org. (2021) Abu Ghraib torture and prisoner abuse. Wikipedia.org. https://en.wikipedia.org/wiki/Abu_Ghraib_torture_and_prisoner_abuse. Published 2021. Accessed 16 Apr 2021
21. Wikipedia.org. Guantanamo Bay detention camp. https://en.wikipedia.org/wiki/Guantanamo_Bay_detention_camp. Accessed 26 Apr 2021
22. Carnahan T, McFarland S (2007) Revisiting the stanford prison experiment: could participant self-selection have led to the cruelty? Personal Soc Psychol Bull 33(5):603–614. https://doi.org/10.1177/0146167206292689
23. Haslam SA, Reicher SD (2012) When prisoners take over the prison: a social psychology of resistance. Personal Soc Psychol Rev 16(2):154–179. https://doi.org/10.1177/1088868311419864
24. Haslam SA, Reicher S (2007) Beyond the Banality of evil: three dynamics of an interactionist social psychology of tyranny. Personal Soc Psychol Bull 33(5):615–622. https://doi.org/10.1177/0146167206298570
25. Möller-Leimkühler AM, Bogerts B (2013) Collective violence—neurobiological, psychosocial and sociological conditions. Nervenarzt 84(11):1345–1354, 1356–1358. https://doi.org/10.1007/s00115-013-3856-y
26. Anderson CA, Bushman BJ (2002) Human aggression. Annu Rev Psychol 53:27–51. https://doi.org/10.1146/annurev.psych.53.100901.135231
27. Allen JJ, Anderson CA, Bushman BJ (2018) The general aggression model. Curr Opin Psychol 19:75–80. https://doi.org/10.1016/j.copsyc.2017.03.034
28. Finkel EJ, Hall AN (2018) The I^3 Model: a metatheoretical framework for understanding aggression. Curr Opin Psychol 19:125–130. https://doi.org/10.1016/j.copsyc.2017.03.013
29. Paulhus DL, Curtis SR, Jones DN (2018) Aggression as a trait: the Dark Tetrad alternative. Curr Opin Psychol 19:88–92. https://doi.org/10.1016/j.copsyc.2017.04.007
30. Fiedler P (2018) Epidemiology and course of personality disorders. Zeitschrift für Psychiatr Psychol und Psychother 66(2):85–94. https://doi.org/10.1024/1661-4747/a000344

Chapter 13
Violence as an End in Itself and Lust Gain

The term hedonism derives from the ancient Greek word *hedone*, which means "joy, pleasure, delight, sensual desire." In view of the immense number of victims of violence as well as the psychological and material long-term damage, one would assume that violence is abhorred by everyone, or at least cannot be experienced as pleasurable, and that violent scenes would therefore have long since disappeared from our everyday lives. However, that this is by no means the case is not only due to the fact that violence often seems justified as a reaction to an attack on one's own personal integrity, but that violence can also have other, namely hedonistic-pleasurable, motives. Experiencing depictions of violence, like actively participating in violent acts, has a stimulating or even fascinating component on a not inconsiderable number of people. Some perpetrators are motivated to commit acts of violence, even to the point of excess, for this reason alone.

Current and Historical Examples

Fascination with violence is present everywhere. This is evidenced by the high ratings on television for crime thrillers, which can be seen at prime time and whose content revolves around murder and manslaughter. Martial arts arenas are well filled with boxing events, wrestling or ultimate fighting (Fig. 13.1). Examples of the fun of active violence are riots by hooligans, ultras and riot tourists, who seek out violent events because of their sensational character in order to prove their own strength and significance. The same applies to the considerable sales of videos and apps with killer games whose goal is to kill as many virtual opponents as possible.

The mass consumption of displays of violence in the media cannot be explained in any other way than by the accompanying hedonistic-stimulating effect on viewers. Nor is it far-fetched, from this perspective, to classify team sports such as soccer as ritualized and regimented group aggression, with the sole aim of making players and fans triumph in victory over the opposing team. How deeply this nature is anchored,

Fig. 13.1 Martial arts: violence display for entertainment. Partial images from Pixabay.com no: 83358, image by: David Mark/2,142,472, image by: Taco Fleur/2,495,816, image by: Mirko Zax/100,733, royalty free images. Available online at: https://pixabay.com/de/; (retrieved October2020)

especially in male psychology and physiology, is shown by the fact that happiness hormones and testosterone levels rise in the winners of sporting competitions [1]. The fascination of tens of thousands who show solidarity with their favorite team and long for victory over the opposing team and thus their own feeling of happiness fills the stadiums.

The Roman emperors already knew that the performance of fighting attracts crowds (soccer did not exist at that time), and they had the Colosseum and amphitheaters built for the performance of animal and gladiator fights in order to excite the people about the imperial offer and thus increase their own popularity.

The first great work of Western literature and, along with the *Odyssey*, the most famous work of antiquity, is Homer's *Iliad*, which takes place during the time of the Trojan War around 1200 BCE and is a theatrical and glorifying depiction of acts of violence and scenes of triumph over slain enemies. Cruel public executions for the entertainment of the masses were as common in the persecution of early Christians as they were in the punishment of rebellious slaves, 6000 of whom were crucified before the eyes of spectators along the Via Appia outside Rome after the suppression of the Spartacus rebellion in 71 BCE [2]. Glorification of violence and enthusiasm for war are well-known phenomena not only from antiquity and the time of the Crusades (called for in 1095 CE by Pope Urban II under the motto "God wills it"), but also from more recent European history. One example is the war euphoria in the weeks before the First World War in Germany, Austria, Hungary and the Ottoman Empire on the one side, as well as in England, France and Russia on the other.

The desire for and pleasure in violence is obviously part of human nature across all cultures and continents as an anthropological constant. It is acted out when the social and political framework conditions make it possible and empathy and humanity lose out.

Torture and Sadism

Extreme forms of hedonistic violence are torture and sadism. Medieval torture methods and instruments of torture, which can still be seen in museums and castle cellars from that time, were used on hundreds of thousands of victims and were found in one form or another in many cultures. Elaborate methods of torture existed not only at the torture stakes of the American Indians, but also in the prisons, dungeons and camps of all dictatorships, where torturers practiced the hedonistic acting out of violence. Torture and executions not so long ago attracted crowds in public form, for whom the agony of the victims was a kind of spectacle. Human sacrifices, also of a cruel kind, were part of the rituals of ancient religious ceremonies as well as of the Aztecs in Central America, where sadistic inclinations were acted out under religious pretexts.

The same is true for witch hunts and witch burnings in the Middle Ages. After two Dominican monks published the *Malleus Maleficarum* (*Witches' Hammer*) in the fifteenth century, in which they called for the burning of women who were in league with Satan and were doing the devil's work, about 60,000 women were burned at the stake after being tortured right up until the eighteenth century [3]. The perpetrators and many of the bystanders believed that this was justified and felt satisfaction in doing so.

The same motives can be attributed to the actors of the Inquisition, through which blasphemers, heretics and apostates were forced to confess in the name of the Church through torture and then eliminated. The Spanish Inquisition alone is said to have tortured several thousand people to death from the end of the fifteenth to the beginning of the eighteenth century [4].

State-mandated or tolerated torture, far from being phenomena of the distant past, is ongoing. Let us remind ourselves of the torture methods in the prisons of Central and South American military dictatorships, of the US and British military prisons after the last Iraq war (2003), [5] and of the atrocities of the so-called Islamic State, to name just a few. Undoubtedly, torture methods were and still are also practiced in many other places from which no information emerges.

Sadistic Serial Killers

There are numerous examples of sadistic serial killers for whom lust for murder, often combined with sexual arousal, was the only motive for often gruesome acts of murder. For example, Volker E., born in 1959, long-distance truck driver and murderer of eight women, mostly prostitutes, who, until his arrest in 2006, first asked his unsuspecting victims to let him tie them up for an extra charge, then strangled them under sexual excitement while fetishistically playing with their long hair [6]. He cut off the hair as a trophy and photographed the victims. During psychiatric examination, he stated that he had felt an irresistible urge to commit such acts and afterwards had felt a relaxed sense of satisfaction and his own greatness. After he had committed suicide in custody, a brain autopsy was performed. Minor tissue damage, probably present from childhood, was found in parts of the frontal brain and limbic system, brain areas responsible for the control of archaic urges [6].

An extreme example of a sadistic serial killer from the Middle Ages is the French knight Gilles de Rais, who kidnapped well over 100 children in the neighborhood of his estate over a period of many years without being found out, hung them up by their arms in his castle, sexually abused them and tortured them to death, sometimes using satanic rituals. He was finally convicted and hanged in 1440 [7].

Among better-known examples from more recent times are Jack the Ripper, who in the fall of 1888 terrified London with a series of gruesome murders of prostitutes, but whose identity has never been determined to this day, and Haarmann, the Butcher of Hannover, who between 1918 and 1924 lured at least 24 young men, mostly hustlers or runaways, into his apartment and killed them by biting their throats during sexual acts, then chopping them up to dispose of the body parts. He ended up under the guillotine in 1925 [8].

Worldwide, across all continents, 450 similar cases are documented in the *Encyclopedia of Serial Killers* [8]. Very often, these were sex murders in which the perpetrators were caught only after many acts, and in some cases not at all. The victims were predominantly children, young women, young men and prostitutes. Sexual sadism was not always the motive, often it was pure addictive murder lust and bloodlust. Sometimes the perpetrators were psychotics driven by delusions. Many serial killers report that just before their murders they had a libidinous feeling of intense tension and restlessness and the need to seek a victim; then the excitement of being able to seize the victim and of carrying out the crime; finally, the sensation of relief and diminishing tension. Such serial offenders even exert a certain fascination over some people.

Revenge

The desire for revenge, to pay like back with like or even worse, is one of the strongest motives for the use of violence. Everyone has experienced more or less intense feelings of revenge. In the civilized world, however, it usually remains as a "clenched fist in the pocket," sometimes even with powerless rage. Today, measures of retribution and atonement are usually left to the judiciary or to the paragraphs of the penal codes.

However, it was quite different in the early days of mankind, as well as both in antiquity and in the chivalric feud system of the Middle Ages, until the emergence of the state monopoly on the use of force in modern times. The number of murders, many of them retaliatory, was 50 times higher in Western Europe then than it is today [9] (see Fig. 3.7).

Blood revenge is one of the main motives for murder everywhere in the world where tribal wars still exist and where a state monopoly on the use of force does not exist or has disappeared. Revenge and retaliation are said to be responsible for 10–20% of deadly acts of violence worldwide. Blood revenge is still practiced today in some regions of the Balkans, Sicily (*vendetta*), the North Caucasus and the Philippines [10].

Revenge and retribution are a combination of reactive and proactive/appetitive violence and are perceived as satisfaction by those who are avenged, if they survived, as well as by their relatives.

Revenge activates the brain's reward system in the same way as other things perceived as pleasurable, such as sweets [11, 12]. Even though revenge can therefore be perceived as sweet, the suppression of the desire for revenge, the punishment of injustice by the judiciary and the possibility of forgiveness are among the essential advances of civilization.

Collective Violence as a State of Ecstasy

Collective violence can take on an intoxicating character. Extensive research on the motivation of cruel acts of violence has been carried out on former child soldiers from the regions of unrest in the east of the Democratic Republic of Congo and Uganda, as well as on former perpetrators of the genocide between Tutsis and Hutus in Rwanda [13–15]. In 1996, bloody civil wars raged in the former Zaire, for which numerous child soldiers were forcibly recruited. Two years earlier, within a few months, between 800,000 and 1,000,000 Tutsis—an average of about 10,000 dead per day—were massacred by Hutus with machetes and clubs after being declared vermin (cockroaches) by the butchers. In the end, 100,000 Hutus were also dead. Many of the combatants interviewed later were not traumatized by the experience of their horrific acts. On the contrary, it turned out that after overcoming initial inhibitions, with the increasing exercise of violent actions, instead of post-traumatic stress disorder, which would have been accompanied by negative affects, such as fear, despair and fright, a kind of hunting instinct with the desire to kill people was awakened. In the process, a sense of stimulation and a kind of fascinating excitement developed in the children and adolescents recruited. These young fighters eventually developed a desire to commit atrocities, to see blood and to kill, which they understood as a pleasant, stimulating task [13, 16–19].

This type of aggression has so far been considered mainly as an expression of pathological personality development. However, the sequences of wars, battles, skirmishes and genocides show that hedonistic aggression, also referred to as appetitive aggression, [19] is one of the main underlying motivations for human violence in general and the willingness to kill in conflict regions. Fascination with bloody violence even reached such dimensions that combatants experienced the sight and smell of blood as stimulating and addictive [16].

Vietnam veterans also reported, as did soldiers of the Wehrmacht in World War II, that violence, destruction and acts of killing can take on intoxicating states [20]. The same applies to many events on and behind the Eastern Front during the Second World War and also to the carnage in the city of Nanking in December 1937, in which between 200,000 and 300,000 Chinese civilians were massacred by Japanese soldiers in a kind of killing frenzy.

The significance and extent of hedonistic violence is illustrated not only by the example of soldiers in special war situations, but also by the immense extent of the use of killer games and ego-shooters (with evocative names such as *Counter-Strike* and *Call of Duty*) by predominantly male youths. The virtual hunting scenes of people that take place in such games are accompanied by a feeling of satisfaction and of one's own significance in the event of killer success.

Collective violence is also experienced as euphoric outside military actions. A remarkable piece of reportage by an author who joined English hooligans in the 1980s [21] describes in detail the ecstatic excitement during fights against other hooligan groups and rioting in the cities of the opposing team. The author, having not previously imagined that such an effect could happen to himself, was caught up in the euphoria of fighting. Once the barrier in the horde had fallen—as he described it—there was a roar and everyone threw themselves into the fight as if gravity had been suspended. Nothing could stop such an unleashed horde, except the physical force of the police. Group violence was described as one of the strongest experiences that gives those who are able to surrender to it one of the most intense sensations of pleasure. The author himself felt carried away into a kind of intoxication and could understand that mass violence can act like a drug. The pure elemental pleasure that mass violence brings is of an intensity that cannot be compared to anything else. But it is not just any kind of violence, such as a profane brawl; it is the violence of masses that is important here, the very special mechanism of violence involving large numbers [21]. Similar descriptions exist from other connoisseurs of the hooligan scene [22].

Hedonistic Violence as a Relic of Phylogenesis

The current and historical examples mentioned above show, impressively, that aggression and violence can occur not only as reactions to provocation, frustration or social exclusion, and also not only as an expression of a lack of moral values education in childhood, a psychological disorder or brain tissue defects. The successful practice can have a hedonistic, or pleasure-oriented, component, just like the perception of violent acts, which is why violence is accepted and practiced for its own sake.

The predisposition for this, or at least traces of it, is present in every human being, as a dark mortgage passed on in phylogenesis, as a kind of collective unconscious, but it not infrequently reaches degrees of expression in which violence is practiced excessively, because of its stimulating, euphoric and fascinating effects.

Hedonistic aggression can be derived from the perspective of evolutionary psychology in the same way as other archaic mental traits, even if it seems counterintuitive at first that pleasure in killing one's own kind can be associated with an evolutionary advantage. The person who felt pleasure in killing others and did so successfully had, first of all, a higher chance of passing on his genes and the accompanying character traits to his offspring than the inferior victim [23]. However, the perpetrator himself ran a high risk of becoming the victim of acts of revenge or, with emerging civilization and jurisprudence, of being hanged on a gallows or locked away. This in turn limited the inheritance of such mentalities and was probably the reason why lust murderers—apart from a few tyrannical dictators—could not gain the upper hand in human history. In today's civilized world, they represent criminological exceptions.

The phylogenetic selection of collective hedonistic aggression is different from individual aggression. The group of early humans who enjoyed the destruction of a competing group and were successful in doing so, celebrated themselves and could pass on this group mentality with their gene pool without having to fear sanctions or revenge by their own group or by the defeated victims. In contrast to individual sadism, therefore, euphoric group aggression has remained ubiquitous throughout world history to this day. Not only the actions of hooligans, but also the recurring enthusiasm for war throughout history, even in intellectual circles, bear witness to this. Team sports like soccer are so attractive because the victory of "one's own" team activates feelings of euphoria—a relic from our evolutionary history. However, hardly any of the fans may be aware of this.

In analogy to human hedonistic violence, appetitive aggression is also found in the animal kingdom. Such aggression occurs, for example, when a predator attacks a prey animal. In case of success, appetitive aggression has a rewarding character and spurs on further prey capture. Anyone who has observed a cat playfully tormenting a captured mouse for a time before eating it knows that an appetitive component of aggression (animal sadism?) occurs in other species as well [23]. The wide distribution of appetitive aggression in the animal world points to a phylogenetic selection of this type of aggression that has been going on for many millions of years, beginning long before the emergence of mankind, and which may have already shaped the behavior of *Tyrannosaurus rex.*

Hunting instincts combined with the feeling of satisfaction at the success of the hunt were essential for the survival of our ancestors in the course of human development. Remnants of this have survived from ancient and medieval hunting events to "diplomat hunts" and can still be found as archaic remnants in parts of today's hunters.

Brain Biological Correlates of Hedonistic Aggression

Considering that appetitive aggression must have been present in evolution long before the appearance of *Homo sapiens* and its hominid predecessors and probably already helped the Cretaceous predatory dinosaurs to survive 100 million years ago as well as today's reptiles descended from them, it is not surprising that the nerve cell groups responsible for appetitive as well as for reactive aggression are located in the phylogenetically oldest regions of our brain [24, 25]. They are located in the brainstem and hypothalamus regions that have a very similar structure and function in all vertebrates (which biologically includes humans) and have therefore been called the "reptilian brain" (see Chap. 6, Figs. 6.3 and 6.4). In animal experiments, it has been demonstrated that reactive, rage-induced violence is evoked by stimulation of the near midline parts of the hypothalamus, which in turn activate brainstem autonomic functions such as pulse rate, blood pressure, respiratory rate and sweating tendency. Proactive-appetitive-hedonistic violence, on the other hand, is triggered when more laterally located parts of the hypothalamus are stimulated, which under physiological conditions are activated by central parts of the amygdala. [26, 27] (see Chap. 6, Fig. 6.2). Proactive-appetitive aggression does not occur suddenly and abruptly after electrical brain stimulation, nor is it accompanied by the vegetative reactions typical of anger, as is the case with reactive violence, but rather by a tense-calm, planned, deliberate attacking action sequence (e.g., ambushing, stalking and overpowering a prey animal).

The central brain biological principle for the occurrence of appetitive aggression is the interaction of the cell groups located in the hypothalamus, whose activation triggers appetitive-aggressive behavior, with the reward system located in the immediate vicinity in the brainstem and limbic structures [26, 27] (see Fig. 13.2).

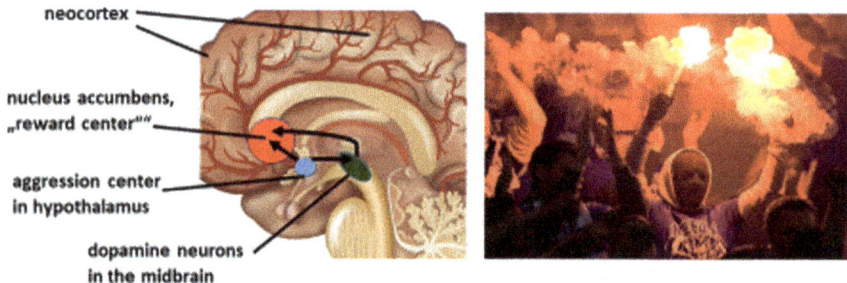

Fig. 13.2 Left: brain areas activated during appetitive/hedonic violence; right: euphoric hooligans. There is a close interaction between nucleus accumbens (red), where dopamine is released from midbrain cell groups (green), and the centers for proactive aggression of the hypothalamus (blue). The same brain systems are activated during all pleasurable experiences as well as by drug use. They are closely interconnected with other structures of the limbic system and the neocortex. Partial image on the left Depositphoto No.: 8610494; partial image on the right Depositphoto No. 130876316. Own labeling

All experiences that are perceived as pleasant and stimulating activate the brain's reward system. The key structures of this system are located in the limbic system, and here again a structure central to reward mechanisms, the nucleus accumbens, plays an important role (Figs. 6.4 and 13.2). This brain area is activated by the neuronal transmitter dopamine, which is produced in cells of the midbrain. All pleasurable activities, such as food intake, sex or even winning money, stimulate the nucleus accumbens by releasing the neurotransmitter dopamine, as do addictive substances (e.g., heroin, cocaine and alcohol). Once the reward center has been stimulated by a certain activity, however, it runs the risk of becoming addicted to such activities and wanting to repeat them constantly or to keep seeking out situations in which this is made possible [23, 25, 28].

The same brain reward system is also activated by hedonistically performed violence or perception of violence [12, 23, 29]. Animal experiments have shown that rewarding effects of aggression in the form of attacking a conspecific are accompanied by activation of the nucleus accumbens and functionally closely related parts of the aggression circuits of the diencephalon and limbic system [27, 30]. Strains of laboratory mice and rats bred for aggressive behavior seek out opportunities to fight opponents, triggering the brain's reward system with the release of the neuronal transmitter dopamine [31]. When a mouse belonging to such a strain bred for aggression is placed in a cage with a foreign mouse, the intruder is immediately attacked. Mice bred for aggression can also be taught to push buttons or levers in their cage to cause them to have another mouse placed in the cage as a reward, which they can then immediately pounce on [25, 32]. The opportunity to attack a foreign mouse and the accompanying release of the rewarding neurotransmitter in the nucleus accumbens has a rewarding and behavior-reinforcing effect on the aggressive animal similar to food, sex or drugs, which activate the same brain mechanisms. Substances that block the action of dopamine in the reward system (neuroleptics) can attenuate this behavior [30].

The astonishing uniformity of structure and function of these phylogenetically ancient brain parts responsible for aggression and reward in all vertebrates—from crocodiles to humans—implies that animal findings on the physiology of these brain systems are transferable to humans. The difference between the human brain and that of the laboratory animal is that humans have a huge neocortex that can keep the archaic emotions of the reptilian brain in check and adapt them more intelligently than the animal to different circumstances. Also, the neocortex has an almost unlimited storage capacity for learned ethical and moral norms and prosocial attitudes that put a stop to the reptilian brain in us—provided that such violence-inhibiting norms and attitudes have been imparted and stored in the neocortex.

A weakened form of sadism is *Schadenfreude*. In an experiment, subjects were shown scenes that evoked *Schadenfreude* in them; brain activity was examined by functional magnetic resonance imaging while they were pleased with the misfortune of others. A portion of the brainstem was activated, including the nucleus accumbens, which is part of the reward system and evokes sensations of pleasure and joy [33].

More often than by hedonistic violence, feelings of satisfaction are elicited by empathy, helpfulness, affection, selflessness and fellow humanity. These prosocial behaviors also activate brain areas that cooperate closely with the reward system [34–36]. If they are not sufficiently encouraged or are eclipsed by violence-glorifying ideologies, appetitive aggression and hedonistic violence, which has been ingrained in our reptilian brain from time immemorial, will again determine the course of history.

References

1. Carré JM, Archer J (2018) Testosterone and human behavior: the role of individual and contextual variables. Curr Opin Psychol 19:149–153. https://doi.org/10.1016/j.copsyc.2017.03.021
2. Wikipedia.org. Third Servile War. Wikipedia.org. https://en.wikipedia.org/wiki/Third_Servile_War. Published 2021. Accessed April 16, 2021
3. Wikipedia.org. Witch trials in the early modern period. https://en.wikipedia.org/wiki/Witch_trials_in_the_early_modern_period#Estimates_of_the_total_number_of_executions. Accessed April 15, 2021
4. Wikipedia.org. Inquisition. https://en.wikipedia.org/wiki/Inquisition. Published 2021. Accessed April 15, 2021
5. Wikipedia.org. Abu Ghraib torture and prisoner abuse. Wikipedia.org. https://en.wikipedia.org/wiki/Abu_Ghraib_torture_and_prisoner_abuse. Published 2021. Accessed April 16, 2021
6. Nedopil N, Blümcke I, Bock H, Bogerts B, Born C, Stübner S (2008) Sadistic fetishism—deadly passion. Nervenarzt 79(11):1249–1262. https://doi.org/10.1007/s00115-008-2562-7
7. Wikipedia.org. Gilles de Rais. Wikipedia.org. https://en.wikipedia.org/wiki/Gilles_de_Rais. Published 2021. Accessed April 16, 2021
8. Murakami P, Murakami J (2003) Lexikon Der Serienmörder. 450 Fallstudien Einer Pathologischen Tötungsart (Encyclopedia of Serial Killers. 450 Case Studies of a Pathological Type of Killing.). München: Ullstein Verlag
9. Pinker S (2011) The better angels of our nature: why violence has declined, Chap. 2. Viking Books Adult
10. Wikipedia.org. Feud. Wikipedia.org. https://en.wikipedia.org/wiki/Feud#Blood_feuds. Published 2021. Accessed April 16, 2021
11. de Quervain DJ-F (2004) The neural basis of altruistic punishment. Science (80-) 305(5688):1254–1258. https://doi.org/10.1126/science.1100735
12. Chester DS, DeWall CN (2016) The pleasure of revenge: retaliatory aggression arises from a neural imbalance toward reward. Soc Cogn Affect Neurosci 11(7):1173–1182. https://doi.org/10.1093/scan/nsv082
13. Hecker T, Hermenau K, Maedl A, Elbert T, Schauer M (2012) Appetitive aggression in former combatants—derived from the ongoing conflict in DR Congo. Int J Law Psychiatry 35(3):244–249. https://doi.org/10.1016/j.ijlp.2012.02.016
14. Elbert T, Schauer M, Hinkel H et al (2013) Sexual and gender-based violence in the Kivu provinces of the democratic Republic of Congo. World Bank, Washington, DC. http://documents1.worldbank.org/curated/en/795261468258873034/pdf/860550WP0Box380LOGiCA0SGBV0DRC0Kivu.pdf
15. Crombach A, Elbert T (2015) Controlling offensive behavior using narrative exposure therapy. Clin Psychol Sci 3(2):270–282. https://doi.org/10.1177/2167702614534239

References

16. Moran JK, Dietrich DR, Elbert T, Pause BM, Kübler L, Weierstall R (2015) The scent of blood: a driver of human behavior? Hoffmann H (ed). PLoS ONE 10(9):1–18. https://doi.org/10.1371/journal.pone.0137777
17. Nandi C, Crombach A, Bambonye M, Elbert T, Weierstall R (2015) Predictors of posttraumatic stress and appetitive aggression in active soldiers and former combatants. Eur J Psychotraumatol 6(1):26553. https://doi.org/10.3402/ejpt.v6.26553
18. Elbert T, Weierstall R, Schauer M (2010) Fascination violence: on mind and brain of man hunters. Eur Arch Psychiatry Clin Neurosci 260(S2):100–105. https://doi.org/10.1007/s00406-010-0144-8
19. Elbert T, Moran JK, Schauer M (2017) Lust for violence: appetitive aggression as a fundamental part of human nature. e-Neuroforum 23(2). https://doi.org/10.1515/nf-2016-A056
20. Neitzel S, Wetzler H (2017) Soldaten—Protokolle Vom Kämfen, Töten Und Sterben [Soldiers—Protocols of Fighting, Killing and Dying]. Fischer-Verlag, Frankfurt a.M.
21. Buford B (1991) Among the thugs: the experience, and the seduction, of crowd violence. Secker & Warburg, London
22. Claus R (2018) Hooligans. Eine Welt Zwischen Fußball, Gewalt Und Politik [A World between Football, Violence and Politics]. Verlag Die Werkstatt GmbH
23. Nell V (2006) Cruelty's rewards: the gratifications of perpetrators and spectators. Behav Brain Sci 29(03):211–224. https://doi.org/10.1017/S0140525X06009058
24. Haller J (2013) The neurobiology of abnormal manifestations of aggression—a review of hypothalamic mechanisms in cats, rodents, and humans. Brain Res Bull 93:97–109. https://doi.org/10.1016/j.brainresbull.2012.10.003
25. Golden SA, Jin M, Shaham Y (2019) Animal models of (or for) aggression reward, addiction, and relapse: behavior and circuits. J Neurosci 39(21):3996–4008. https://doi.org/10.1523/JNEUROSCI.0151-19.2019
26. Flanigan ME, Russo SJ (2019) Recent advances in the study of aggression. Neuropsychopharmacology 44(2):241–244. https://doi.org/10.1038/s41386-018-0226-2
27. Falkner AL, Grosenick L, Davidson TJ, Deisseroth K, Lin D (2016) Hypothalamic control of male aggression-seeking behavior. Nat Neurosci 19(4):596–604. https://doi.org/10.1038/nn.4264
28. Kareken DA (2018) Missing motoric manipulations: rethinking the imaging of the ventral striatum and dopamine in human reward. Brain Imaging Behav 13:306–313. https://doi.org/10.1007/s11682-017-9822-8
29. Urban NBL, Slifstein M, Meda S et al (2012) Imaging human reward processing with positron emission tomography and functional magnetic resonance imaging. Psychopharmacology 221(1):67–77. https://doi.org/10.1007/s00213-011-2543-6
30. Couppis MH, Kennedy CH (2008) The rewarding effect of aggression is reduced by nucleus accumbens dopamine receptor antagonism in mice. Psychopharmacology 197(3):449–456. https://doi.org/10.1007/s00213-007-1054-y
31. May ME, Kennedy CH (2009) Aggression as positive reinforcement in mice under various ratio- and time-based reinforcement schedules. J Exp Anal Behav 91(2):185–196. https://doi.org/10.1901/jeab.2009.91-185
32. Covington III HE, Newman EL, Leonard MZ, Miczek KA (2019) Translational models of adaptive and excessive fighting: an emerging role for neural circuits in pathological aggression. F1000Research 8:963. https://doi.org/10.12688/f1000research.18883.1
33. Takahashi H, Kato M, Matsuura M, Mobbs D, Suhara T, Okubo Y (2009) When Your gain is my pain and your pain is my gain: neural correlates of envy and schadenfreude. Science (80-)323(5916):937–939. https://doi.org/10.1126/science.1165604
34. Bernhardt BC, Singer T (2012) The Neural Basis of Empathy. Annu Rev Neurosci 35(1):1–23. https://doi.org/10.1146/annurev-neuro-062111-150536
35. Singer T, Klimecki OM (2014) Empathy and compassion. Curr Biol 24(18):R875–R878. https://doi.org/10.1016/j.cub.2014.06.054
36. Kanske P, Böckler A, Trautwein F-M, Singer T (2015) Dissecting the social brain: introducing the EmpaToM to reveal distinct neural networks and brain–behavior relations for empathy and theory of mind. Neuroimage 122:6–19. https://doi.org/10.1016/j.neuroimage.2015.07.082

Chapter 14
Social Causes of Violence

B. Bogerts and C. Steinmetz

In addition to genetic, brain-biological and biographical constellations, the current social environment, group-dynamic effects and the overall social situation play a decisive role in the multidimensional causal structure of violence. This is the explanation of why the extent of violence varies so much throughout history and in different regions of the world.

Historical and Geographical Variations in the Frequency of Violence

An individual's propensity to violence depends, on the one hand, on their inherited personality traits, which are modified by early family influences and by experiences in the course of life history, and, on the other hand, on the influences that are present in the current situation and that provoke or prevent violence. Moreover, it is also significantly shaped by the long-term social and political environment. The latter explains why murder rates in some countries of Central America are more than 50 times higher than in Western Europe and also why the frequency of manslaughter and murder in the Middle Ages in Europe was many times higher than today. Another impressive example of the importance of political, social and cultural influences on the propensity for violence is recent German and Japanese history; both countries were largely militaristic states in the first half of the twentieth century, with many violent conflicts which also instigated world wars; today they are perhaps among the most peace-loving. The rate of violence is thus subject to considerable regional and temporal variation, although there is little difference in the basic mental make-up of the average persons over the centuries or between different countries and continents. How dependent the rate of violence is on the political and social system is shown by the development of the number of murders in two regions that were previously British colonies and thus had comparable administrative structures before they became independent; these are Jamaica and Singapore (Fig. 14.1).

© The Author(s), under exclusive license to Springer Nature Switzerland AG 2021
B. Bogerts, *Where Does Violence Come From?*,
https://doi.org/10.1007/978-3-030-81792-3_14

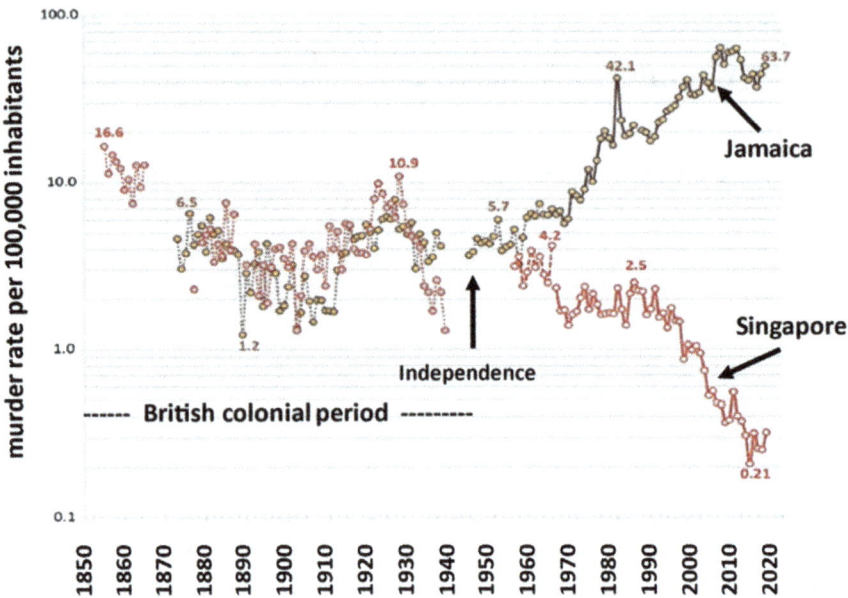

Fig. 14.1 Development of homicide rates in Jamaica and Singapore after the end of the British colonial era. *Source:* Adapted from Global Study on Homicide 2019, Executive Summary, p. 28 [1]

Until independence in the 1960s, the murder rate was about the same in both countries. Thereafter, by 2016, it had risen to one of the world's highest in Jamaica (64/100,000) and dropped to the world's lowest in Singapore (0.21/100,00); the rate is thus about 300 times higher in Jamaica today than in Singapore [1].

In Singapore, the end of colonialism was followed by an intensive strengthening of the educational, social and health care systems, the fight against corruption and the establishment of a functioning judiciary. In Jamaica, there were ongoing feuds between rival political parties, bribery, increased gang crime, easy access to firearms and weak rule of law with an insufficiently established state monopoly on the use of force [1].

Quite obviously, the extent to which the readiness for reactive or appetitive individual or collective violence anchored in our reptilian brain is inhibited or can unfold unhindered depends on the social framework conditions.

The Importance of the State Monopoly on the Use of Force

The frequency of violence in different regions of the world and historical periods depends crucially on an effective state monopoly on the use of force. This has been the subject of numerous social science discussions. However, with regard to the manifestations and causes of violence, social scientists have so far mainly dealt with power, domination and order, and less with the conditioning factors for the occurrence of physical violence by individuals or groups. The state monopoly on the use of force has been defined from the point of view of social science as a suitable means of attaining and maintaining power. According to the German sociologist Max Weber (1864–1920), the state is defined as a "human community that (successfully) claims the monopoly of the legitimate use of physical force within a given territory." [2].

In fact, it was the introduction of the state's monopoly on the use of force that put an end to the ancient and medieval law of the fist and feud system, which demanded many victims and made the use of violence on one's own initiative, whether out of revenge, retaliation, or the desire for dominance or possession, a punishable offense—with the exception of self-defense. Thus, it is ultimately the fear of accusation, condemnation and punishment that enters into the balancing of the consequences of an act of retaliation and makes people shy away from the personal, non-legitimized use of violence, as well as the expectation that injustice will be punished by means of justice. The formation of relatively stable monopolies of violence in the course of the consolidation of the European nations created the basis for violence to lose its commonplace nature and for today's relative poverty of violence in civilized countries to become an everyday experience.

Where the state monopoly on the use of force and the judiciary are functioning and vigilante justice is punished, it does not occur as a rule. Where neither is the case, the incidence of individual and group violence is correspondingly high; at present, this is particularly evident in some states in Africa, Latin America and the Middle East. Warlords, drug barons, militia chiefs, gangster bosses and mafiosi set the rules here.

It was even assumed that interpersonal violence—violence between private individuals—would become increasingly taboo and ultimately disappear altogether. However, this assumption proved to be unrealistic, since the predisposition to exercise violence remains present in every human being—albeit unconsciously in most of us—and under sufficiently high stress or provocation a deficient behavioral control can become apparent in vulnerable personality structures. Violence can never completely disappear from a society, since it is constantly available as an "everyman" resource [3]. Moreover, legitimized state violence used as a last resort is a necessary condition for maintaining social order.

The extent to which the state's monopoly on the use of force can affect violent crime is shown in Fig. 14.2a. It shows data from studies conducted by the Bertelsmann Foundation in 2020 on developing and transition countries (Bertelsmann Transformation Index [4, 5], not including Western industrialized nations). The figure shows a highly significant inversely proportional relationship between the degree of intactness of the state monopoly on the use of force and the incidence of violence within the respective country. This means that within the countries studied, the better the state monopoly on the use of force functions, the lower is the incidence of violent crime.

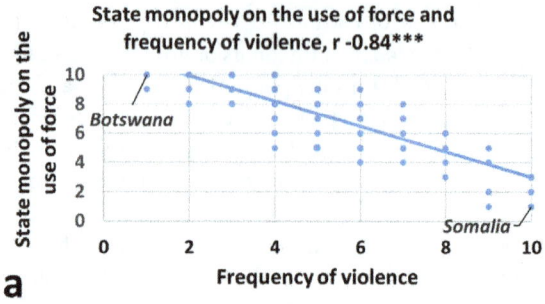
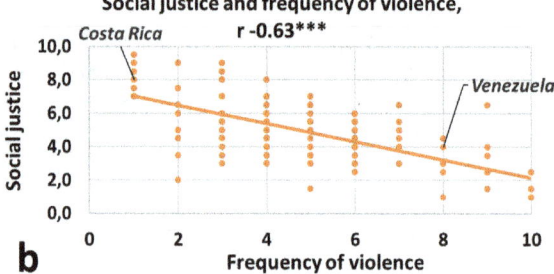
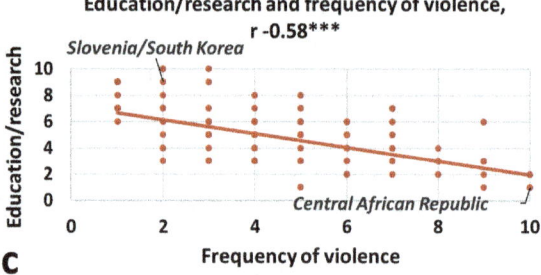
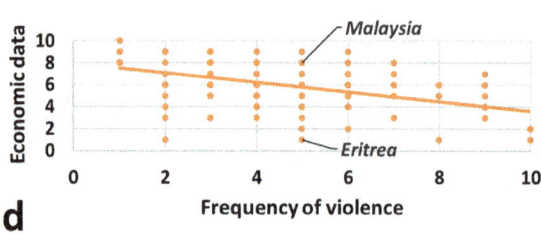

◀**Fig. 14.2** Relationship between indicators of a country's social situation and incidents of violence Data from the Bertelsmann Transformation Index [4], Each point of the diagram represents a country. **a** State monopoly on violence, **b** social justice, **c** education, and **d** economic performance, as well as incidence of violence are divided into levels from 1 to 10 depending on their effectiveness/frequency: 1 means very low, 10 means very high; r = strength of correlation; $r = 1$ means best possible correlation; and $r = 0$ no correlation. *** = correlation is highly significant. The strongest correlation ($r = 0.84$) is between a well-functioning state monopoly on the use of force and a low incidence of violence. Examples from Africa include (Fig. a, top left) Botswana, with a well-functioning state monopoly on the use of force and a low incidence of violent crime, and Somalia, with a largely collapsed state order and a very high murder rate. The weakest correlation is between economic data and incidence of violence (r-0.48, Fig. d, bottom right), although this correlation is also significant. The individual values used to calculate the correlation are subject to a considerable spread between the individual countries; for example, Malaysia and Eritrea have very different economic performance while the incidence of violence is roughly the same (Fig. d, bottom right). This means that factors other than economic performance must play a significant role. Own illustration. *Data source* Bertelsmann Transformation Index; see Ref. [4]

The Downsides of the State Monopoly on the Use of Force

Thomas Hobbes (1588–1679) developed his ideas of the state monopoly on the use of force, the *Leviathan*, out of his experiences in the English Civil War (1642–1649). According to the Hobbes model, citizens, once subject to the state, no longer have control over the Leviathan. The risk of falling to a corrupt and selfish ruler seemed less to Hobbes than a separation of powers, as proposed by the philosophers Locke (1662–1704), Montesquieu (1689–1755), and Kant (1724–1804). Therefore, Hobbes' Leviathan became an argument often used to justify totalitarian state conceptions.

There are numerous examples in history as well as in the present of downsides of the state monopoly on the use of force, some of which reached apocalyptic proportions: National Socialism and the Holocaust, Stalinism, Maoism, Pol Pot and the Khmer Rouge, to name just a few examples from recent history with a combined death toll of more than 100 million (see Chap. 18, Table 18.1). This was only possible because the respective regimes, commanded by power-obsessed and sometimes paranoid rulers, lacked democratic legitimacy and control, or this was eliminated after initial democratic elections.

There are numerous other examples of the state's monopoly on the use of force getting out of hand in the form of paramilitaries, death squads and military forces violently persecuting their own people with the acquiescence of the rulers, usually without being sanctioned.

The emergence of violence is successfully suppressed by state force in order to maintain order and peace only if it is subject to functioning democratic-parliamentary control with party diversity. Otherwise, the risk of abuse by dictators, oligarchs or political systems based on single parties is considerable.

Not only physical use of force, but also systemic oppression is a form of state abuse of power. The term "structural violence" [6] is intended to characterize all those conditions that are responsible for preventing people from developing their potential. This term is intended to point out that malnutrition of large parts of the population and a lack of health care represent such severe damage that a concept of violence that does not address these phenomena is useless [6]. Accordingly, it has been argued that all actions aimed at reducing social participation or causing material disadvantage represent categories of the exercise of power that are comparable to physical violence and can replace it [6].

Police Violence

It is not only the holders of state power, but also the executors of the state's monopoly on the use of force on the ground, in other words, the police or comparable security forces, who sometimes show abuse of power in certain situations. In most countries, it can be assumed that the vast majority of police officers adhere to clearly defined regulations on the use of physical force. However, under very demanding conditions, some of these law enforcement officers may lose control over the archaic mechanisms of impulsive-reactive and also proactive-appetitive aggression that are anchored in every brain. It should be noted that above a certain level of stress, aggression control fails in most average brains.

Certainly, there are some individuals among them who, assuming that their actions have no consequences and in situations that seem suitable for this purpose, act out extremist, racist or simply hedonistic desires for violence. Current events indicate that there seem to be considerable differences between the US and Europe in this respect [7]. Spotting such characters in time during training is not likely to be easy. If such indiscriminate police violence is accidentally filmed by bystanders and given extensive media coverage, understandable mass protests regularly follow, within which not only peaceful demonstrators but also some hooligans who in turn act out hedonistic desires for violence, are to be found.

Economic Conditions and Violence

The risk of violence is higher in countries with low socioeconomic status, low average income, low economic growth and unequal income distribution. Economic and social downturns in a community are associated with increased incidence of violence, the emergence of violent gangs and other forms of organized crime [8]. Societies with highly unequal distribution of income and wealth have a higher incidence not only of mental and physical illness, but also of increased bullying in schools, homicides and terror by their own populations.

As shown in Fig. 14.2d, however, the relationship between a flourishing economy and declining violent crime is rather weak. Although low economic performance and conflict intensity are statistically significantly correlated (i.e., the lower the economic performance, the higher the rate of violence), there are significant differences between countries with comparable economic status. This shows that, in addition to the economic situation, there are a number of other social and person-related social factors that shape the incidence of violence in a society.

Even in historical retrospect, economic prosperity is not always accompanied by a low incidence of violence. In the US, for example, the murder rate fell sharply in the midst of the Great Depression, rose sharply during the boom of the 1960s and reached new lows during the Great Recession that began in 2007 [9]. Violence is more likely to be caused by emotional, personal and biographical than economic factors.

Different aspect from the overall economic situation of a region is the distribution of resources as well as the social security of the population. When resources are unequally distributed, individuals at the lower end of the distribution have little to lose and much to gain. Risky, dangerous behavior, including violent crime, then becomes more tempting.

Therefore, clear correlations are found between economic inequality, which is a reflection of social justice, and violent crime (Fig. 14.2b). For example, a comparison of over 77 neighborhoods in Chicago found a significant association between income inequality and homicide rates [10]. Families with low socioeconomic status, who suffer from a range of disadvantages and stressors, have more than twice the rate of violence than other families [11].

A clear link has also been demonstrated between unemployment and various forms of violence. In both high-income and low-income countries, male unemployment is one of many risk factors for the occurrence of violence [12].

It is not so much the wealth of a society that guarantees more peaceful coexistence, but rather the distribution of this wealth within society, equal opportunities and social security.

Societal Attitudes Toward Violence

The view of how violence is dealt with within a society is subject to a constant process of historical change, which is still ongoing and differs considerably between different cultures.

Violence against women goes hand in hand with unequal rights of the sexes. In some cultures, there is a traditional view that women are subject to the authority of men, which makes women and girls vulnerable to physical, emotional and sexual violence. Countries with pronounced gender inequality have higher rates of physical and sexual interpersonal violence [13].

Under the Sharia law in force in Islamic countries, corporal punishments such as beating, lashing, amputation of limbs, and stoning or other forms of capital punishment are still possible today.

While the need for a state monopoly on the use of force was already generally recognized at the beginning of the twentieth century, it was still considered normal in

Central Europe for violence to be used within the family against women and children for the purpose of corporal punishment. The same applied to members of the servant lower classes or inhabitants of European colonies in other continents. [14] The nonviolent intercourse with family members and subordinates is the result of decades of rethinking. In Germany, for example, it took until 1997 for rape within marriage to be included in the penal code as a criminal offense [15]; parental corporal punishment was not abolished until 2000; and it still exists in most regions of the world.

Anomie and Disintegration as Causes of Violence

The term "anomie" was coined by the French sociologist Émile Durkheim (1859–1917) [16]. Anomie represents a state in which the rules and norms of society no longer apply and, accordingly, violence and criminal acts become relevant options. Durkheim locates this lawless state of anomie above all within processes of social upheaval, since here a social change takes place within a short time, through which previously learned norms and values lose their validity while at the same time the social processes and institutions responsible for the implementation of new social norms have not yet taken effect [17]. The upheaval process taking place during Durkheim's lifetime, in which he observed this state of anomie, was the change from an agrarian to an industrial society (Fig. 14.3).

However, an increase in the death rate due to waves of violence during the process of upheaval could also be observed in later state upheavals, such as after the First World War in Germany during the founding phase of the Weimar Republic [18], and also after the dissolution of the Soviet Union [19] (Fig. 14.4).

The concept of anomie has been expanded to the understanding that it can occur not only within processes of social upheaval, but also when members of disadvantaged social classes or groups are no longer able to achieve their desired social goals by legal means [20, 21]. This theory thus provides an explanatory approach to the usually higher crime rate in poorer neighborhoods, which are disadvantaged due, for example, to low levels of education and the resulting poor career opportunities. The values and norms of a society that disadvantages them are no longer accepted by those affected, which in turn fosters the emergence of anomic conditions in which violence also becomes an option.

A similar perspective is drawn by the social disintegration theory [22–24], according to which various social conditions must be met for people to feel integrated into society. These include sufficient access to labor, housing and consumer markets, educational opportunities and cultural offerings. In addition, there is a personal dimension through the connection of people on a personal level via friendship or as colleagues. If people feel integrated in all areas of society, it can be assumed that they accept the norms of this society. If not, the risk of conflict and violence increases.

Emil Durkheim
(1858-1917)

Fig. 14.3 Émile Durkheim (1858–1917). Émile Durkheim.jpg. Public domain. Available online at: https://de.wikipedia.org/wiki/Émile_Durkheim#/media/Datei (retrieved October 2020)

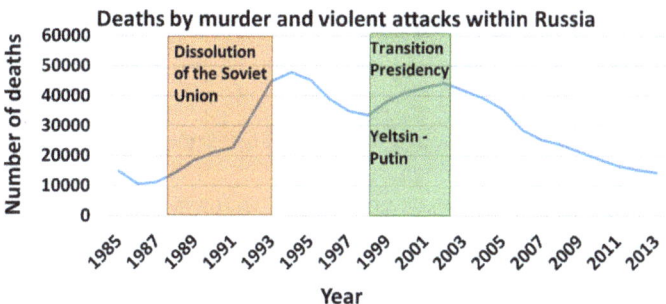

Fig. 14.4 Fatalities from violent crime in Russia during periods of political upheaval from 1985 to 2013. Partial aspects of anomie with weakedning of the state monopoly on the use of force explain the temporary rise in homicide rates at the dissolution of the Soviet Union and the end of Yeltsin's presidency. *Source* Own graph, data from the European Health Information Gateway [19]

The disintegration and anomie theories provide an explanatory framework for the frequency of criminal acts in certain regions and neighborhoods. In a study published as early as 1942, the housing situation of 60,000 delinquent youths in Chicago was investigated [25]. Based on the case studies and statistical analyses conducted, a more frequent occurrence of crime was found in so-called "transition zones," which were located between the industrial core and the "bacon belt" of the city. These areas were characterized by a heterogeneous population structure, low socioeconomic status, and unstable family and housing conditions. At the same time, these zones were subject to constant processes of change due to high population fluctuation, which is why traditionally supportive and regulating institutions such as neighborhoods, schools and families were unable to gain a foothold [26]. The confluence of such factors led to a lack of binding values, a lack of consensus on norms and moral disorientation. Instead, especially among the young, an orientation towards the criminal behavior of other neighborhood residents took place; crime and violence levels increased. Even this early study points to a complex set of conditions that give rise to so-called problem neighborhoods with increased levels of violence. The simplistic assumption that poverty as such or a certain ethnic composition of the residents of a neighborhood lead to more violence had to be replaced by a more differentiated view.

Seventy years after the publication of the Chicago study, similar explanatory patterns for the increased propensity to violence among young people can be demonstrated in European problem neighborhoods. Anomic disorientation and a lack of prospects in connection with the dimensions of social disintegration described above are strong indicators for the emergence of violence [27].

Interaction of Social, Biographical and Neurobiological Conditions

Depending on the type of society and region, very different factors can come together in the complex set of social conditions of violence, which in turn encounter very different personality constellations and mentalities of the actors (Fig. 14.5). This explains the diversity of violent conflicts worldwide. Not all individuals affected by a social situation conducive to violence use violent methods, no matter how justified this may seem. A biographically derivable or neurobiological disposition for this must be present. The latter is predominantly found in the male gender [28].

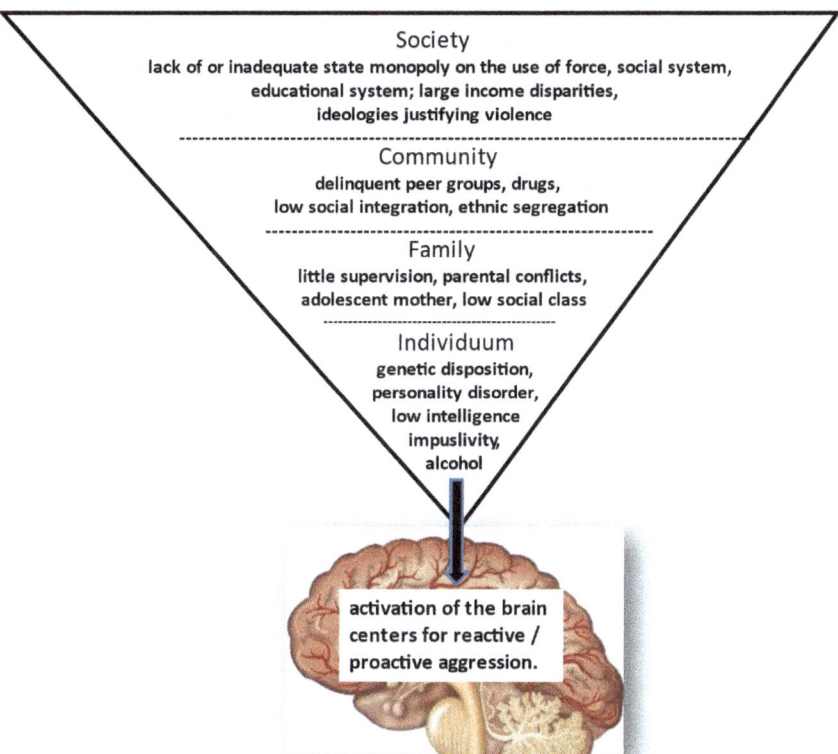

Fig. 14.5 Schematic summary of the multilayered social cause structure of aggression and violence and its influence on brain biology. (*Source* Adapted from Möller-Leimkühler and Bogerts (2013) [28]). The figure is intended to illustrate that the risk factors for individual or collective acts of violence include both the overall social situation and the close social environment (community, family) as well as the individual brain-biological disposition. In the case of situational triggers, this activates the phylogenetically old neuronal generators of violence in the limbic system and brain stem. Partial image from Deposit photos No: 86104948

References

1. UNODC United Nations Office on Drugs and Crime (2019) Global study on homicide. Vienna. https://www.unodc.org/unodc/en/data-and-analysis/global-study-on-homicide.html
2. Munro A (2021) State monopoly on violence; political science and sociology. britannica.com. https://www.britannica.com/topic/state-monopoly-on-violence. Published 2021. Accessed 21 Apr 2021
3. Popitz H (2017) Phenomena of power; authority, domination and violence. In: Poggi G (ed). Columbia University Press
4. BertelsmannStiftung (2020) BTI Bertelsmann transformation index, scores. bti-project.org. https://www.bti-project.org/en/meta/downloads.html. Published 2020. Accessed 19 Apr 2021
5. BertelsmannStiftung (2020) Methodology bertelsmann transformation index. bti-project.org. https://bti-project.org/en/methodology.html. Published 2020. Accessed 19 Apr 2021
6. Galtung J (1969) Violence, peace, and peace research. J Peace Res 6(3):167–191. https://doi.org/10.1177/002234336900600301

7. Wikipedia.org. Police brutality in the United States. Wikipedia.org. https://en.wikipedia.org/wiki/Police_brutality_in_the_United_States#:~:text=In 2019%2C 1%2C004 people were,combat police corruption and brutality. Published 2021. Accessed 19 Apr 2021
8. Long SJ, Fone D, Gartner A, Bellis MA (2016) Demographic and socioeconomic inequalities in the risk of emergency hospital admission for violence: cross-sectional analysis of a national database in Wales. BMJ Open 6(8):1–10. https://doi.org/10.1136/bmjopen-2016-011169
9. Pinker S (2011) The better angels of our nature: why violence has declined; Chap 10. Viking Books Adult
10. Wilson M, Daly M (1997) Life expectancy, economic inequality, homicide, and reproductive timing in Chicago neighbourhoods. Br Med J 314(7089):1271–1274. https://doi.org/10.1136/bmj.314.7089.1271
11. Deegener G, Körner W (2006) Risikoerfassung Bei Kindesmisshandlung Und Vernachlässigung; Theorie, Praxis, Materialien [Risk assessment in child abuse and neglect; theory, practice, materials]. Pabst Science Publisher Verlag, Lengerich
12. WHO (2017) 10 facts about violence prevention. http://www.who.int/features/factfiles/violence/en/. Published 2017
13. Heise LL, Kotsadam A (2015) Cross-national and multilevel correlates of partner violence: an analysis of data from population-based surveys. Lancet Glob Heal 3(6):e332–e340. https://doi.org/10.1016/S2214-109X(15)00013-3
14. Koloma BT (2017) (Staats-)Gewalt und Moderne Gesellschaft, Der Mythos vom Verschwinden der Gewalt [(State) violence and modern society, the myth of the disappearance of violence]. Polit Zeitgesch. 04(2017):16–21
15. Wikipedia.org. Rape in Germany. Wikipedia.org. https://en.wikipedia.org/wiki/Rape_in_Germany. Published 2021. Accessed 20 Apr 2021
16. Durkheim E (1895) Les Règles de La Méthode Sociologique. Alcan, Paris
17. Tierney J (2009) Key perspectives in criminology, vol 1. Open University Press McGraw-Hill Education, New York
18. Jones M (2016) Founding Weimar. Violence and the German revolution 1918–1919. Cambridge University Press
19. World Health Organization RO for E. European Health Information Gateway. gateway.euro.who.int. https://gateway.euro.who.int/en/hfa-explorer/. Published 2020
20. Wikipedia.org. Strain Theory (Sociology). Wikipedia.org. https://en.wikipedia.org/wiki/Strain_theory_(sociology). Published 2021. Accessed April 20, 2021.
21. Wikipedia.org. Social Disorganization Theory. Wikipedia.org. https://en.wikipedia.org/wiki/Social_disorganization_theory. Published 2021. Accessed 20 Apr 2021
22. Wikipedia.org. Wilhelm H (2021) Wikipedia.org. https://en.wikipedia.org/wiki/Wilhelm_Heitmeyer. Published 2021. Accessed 21 Apr 2021
23. Heitmeyer W (2010) What divides Society—what holds society together? In Phenomena of violence—structures, developments and need for responses BKA autumn conference 19–20 Oct 2010. https://www.google.com/url?sa=t&rct=j&q=&esrc=s&source=web&cd=&ved=2ahUKEwjBw83Q44zwAhX__7sIHcPMCe0QFjAIegQIDxAD&url=https%3A%2F%2Fwww.bka.de%2FSharedDocs%2FDownloads%2FEN%2FPublications%2FAutumnConferences%2F2010%2FherbsttagungAutumn2010heitmeyerAbstractAutumn. Published 2010. Accessed 21 Apr 2021
24. Heitmeyer W, Anhut R (2008) Disintegration, recognition, and violence: a theoretical perspective. New directions for youth development. https://www.researchgate.net/publication/23319333_Disintegration_recognition_and_violence_A_theoretical_perspective. Published 2008. Accessed 21 Apr 2021
25. Sanders WB (1943) Juvenile delinquency and urban areas. By Clifford RS, Henry DM, Concerning Juvenile delinquency. By Henry W, Thurston and young people in the courts of New York State. By the Joint Legislative Committee to Investigate Children's Court Jurisd. Soc Forces 21(4):487–488. https://doi.org/10.2307/2571192

26. Ministry_of_Children_Community_and_Social_Services (2021) Social disorganization theory; review of the roots of youth violence: literature reviews; vol 5, Chap 4. children.gov.on.ca. http://www.children.gov.on.ca/htdocs/English/professionals/oyap/roots/volume5/chapter04_social_disorganization.aspx. Published 2021. Accessed 21 Apr 2021
27. Heitmeyer W, Thoma H et al (2012) Gewalt in Öffentlichen Räumen, Zum Einfluss von Bevölkerungs- Und Siedlungssturen in Städtischen Wohnquartieren. [Violence in Public Spaces, On the Influence of Population and Settlement Structures in Urban Neighborhoods], vol 2. Springer Fachmedien, Wiesbaden
28. Möller-Leimkühler AM, Bogerts B (2013) Collective violence—neurobiological, psychosocial and sociological conditions. Nervenarzt 84(11):1345–1354, 1356–1358. https://doi.org/10.1007/s00115-013-3856-y

Chapter 15
Violence in Children and Adolescents: Early Risk Factors

Violence rarely occurs spontaneously for the first time in adulthood, and when it does, it is usually a symptom of a brain disease, a mental disorder or massive stressors. More often, violence is the product of a lifelong development, the most formative stages of which are in childhood and adolescence, when the brain and personality are maturing.

The extent of child and youth violence is subject to considerable fluctuations depending on the social, structural and cultural conditions of a society and has been seen as a reflection of how repressive that society is towards this age group [1]. This following chapter is predominantly dedicated to brain-biological, psychological and clinical aspects.

Neurobiology of the Maturing Brain

The child and adolescent brain is not fully mature until young adulthood (around age 20). Although the nerve cells are all present at birth and the brain already has about 80% of its adult volume by age two, [2] during childhood and adolescence the nerve fibers and their sheaths, and thus the connections between the individual brain areas, grow only gradually. The areas in the brain that are relevant to movement and emotion mature earlier than the phylogenetically newer cortical areas. As the last region, the frontal brain reaches full maturity of its pathway connections and thus full capacity in young adulthood. The frontal brain is the most important instance of planning, deliberative and anticipatory behavior. Therefore, these highest mental abilities are not fully achieved until early adulthood, whereas emotions and agility are already running at full speed during adolescence. This explains why higher impulsivity, directness and spontaneity, and unbridled emotionality are typical of childlike and adolescent behavior. Growing up requires a mature and fully functional frontal brain with sufficient storage capacity and integration ability for life experience, social norms, ethics and morals.

Occurrence and Frequency

The still insufficient maturation of impulse-controlling and anticipatory brain areas of the frontal brain is the reason why aggressive behavior occurs more frequently in childhood than in adulthood. While such behavior is found in 3% of males and 1% of females in the general population, the frequency rates of social behavior disorders in childhood are reported to be about 10% of boys and 5% of girls [3, 4]. The most aggressive phase of life is not found in youths, adolescents or adults but in young children in the age of defiance at about two to three years of age [5]. The tendency to aggressiveness thus seems to be innate and declines in the course of early socialization, provided that social rules and norms are taught and internalized.

Scuffles among children and adolescents are usually not to be classified as malicious-aggressive behavior, but are about the age-typical question of who is the stronger and about clarifying the hierarchy. Often it is just harmless roughhousing or testing of boundaries, which is limited to episodic behavior in the vast majority of adolescents at this age level. More serious and malicious is cyber-bullying via smart phones or tablets, where the perpetrators can remain anonymous [6]. According to a 2017 UNICEF study, two-thirds of children and adolescents reported that they had been victims of bullying [7].

There is a small number of intensive offenders among the violent adolescents, who commit more than half of all thefts, damage to property and bodily harm [8]. Intensive dissocial long-term behavior associated with aggressiveness is shown by about 6% of children and adolescents. Very pronounced disorders of social behavior of clinical relevance were recorded in about 3% of the children, three times more frequently in boys than in girls [4].

Studies on the long-term course of aggressive behavior in childhood showed that in about two-thirds of boys there was no physically aggressive behavior, in about 30%, moderate aggressive behavior, and in 6–7% long-term violent behavior could be observed. This is significantly more frequent and more pronounced in boys than in girls [8]. Bullying, in forms ranging from repeated harassment and defamation to physical attacks, shows a similar frequency [6].

Relationally aggressive behavior, by which is meant talking badly about others, exclusion and bullying, on the other hand, is more frequent among girls [9]. If this behavior is included, aggressive behavior is about equally frequent in boys and girls. However, boys show more visible physical assaults, while girls are more likely to use covert aggression. This difference between boys and girls can already be observed at kindergarten age [5].

Causes of Aggression in Childhood and Adolescence

Early Traumatization

Psychological traumatization in early childhood due to emotional neglect, abuse or experience of violence are among the strongest risk factors for violent acts in adolescence as well as in adulthood. Children with traumatizing early experiences are much more likely to engage in violent acts later in life than children without such experiences [10].

In this context, it is also worth mentioning child soldiers who are used for military purposes and whose number has been estimated at approximately 250,000 in 20 countries around the world, mostly in Africa [11]. These children are sent to combat missions against their will, and in some cases they do so to escape poverty or for revenge. Many such children (but not all, see Chap. 13), if they survive their deployments, bear massive permanent psychological damage and retain a high propensity to violence themselves for life [12].

Multiple delinquent youths are typically exposed to early family and social deficiencies, have experienced violence within the family or even toward themselves, are exposed to social exclusion, and have school problems. Frequently, early alcohol or drug use is also present [13]. Children from a dysfunctional family environment are particularly at risk if they have additional mental deficits due to mild brain dysfunction [3, 14].

Hereditary Factors

In addition to early family and social risk factors, hereditary factors also play a significant role in predisposing to violence in childhood and adolescence. About half of the spectrum of causes for aggressive-dissocial behavior can be traced back to a genetic predisposition in children and adolescents as well as in adults. In particular, violent tendencies associated with callous behavior have a clear genetic component. Such children are unable to perceive the needs and emotions of others [15–17].

That genes play a role is also evident from the fact that boys are more aggressive than girls as early as 17 months of age. Learned social role behavior can hardly be relevant at this age. As in adults, the male Y chromosome obviously already contributes to the gender difference in aggressive behavior in infants.

Mental Disorders: ADHD

Many incarcerated youths exhibit clinically relevant mental disorders [18]. These include, in particular, attention-deficit/hyperactivity disorder (ADHD) and early-onset antisocial personality disorder. More than half of adolescents in prison used alcohol and/or drugs [18]. A meta-analysis found a 21-fold higher risk for aggressive behavior in children with ADHD [19]. Children who show aggressive-dissocial symptomatology early in the development of ADHD require special attention as a potential high-risk group for long-term delinquency [20, 21]. With the help of psychotherapeutic and medication measures, ADHD can be treated well in the vast majority of cases, thus significantly improving the prognosis.

Brain Development Disorders: Toxic Influences

In addition to early childhood psychological trauma and genetic influences, brain developmental disorders and prenatal toxic influences are relevant. There are studies that indicate that children of mothers who smoke during pregnancy are three times more likely to become violent offenders as adults than children without such risks [22]. These children also have a reduction in head circumference and the cerebral cortex in the frontal region is thinner. This indicates that smoking during pregnancy causes impaired brain development in the affected children [22]. The damaged brain regions, especially the frontal brain, have a key role in inhibiting violent behavior.

Another toxic factor that impairs brain development is alcohol. Fetal alcohol syndrome is associated with an increased risk of later inclination to violence [23]. Malnutrition during pregnancy and infancy also impairs brain growth and doubles the risk of antisocial behavior in adolescence. One of the consequences of the famine winter in Holland toward the end of the Second World War in 1944–1945 was that children who were prenatally affected by food shortages during pregnancy were two to three times more likely to have antisocial personality disorder than adults [24].

It is also worth noting that studies have shown that low levels of zinc in the blood are associated with an increased risk of aggressiveness [25]. Zinc is an important trace element for the production of neuronal transmitter substances. The neurons of limbic brain areas relevant to emotions, such as the amygdala and hippocampus, have particularly high zinc content.

A deficiency of the essential amino acid tryptophan has a similar effect. Tryptophan is used to synthesize the neurotransmitter serotonin, which contributes to mental balance. When the tryptophan concentration is lowered in test subjects, they react more aggressively to provocation. When the tryptophan level is increased and the serotonin effect at the synapses is improved by certain antidepressants, aggressiveness decreases again [26, 27].

Of only historical interest now are studies that showed that high lead concentrations had a negative effect on brain development [28]. Lead was used in gasoline and paint until the 1970s. Boys with high blood lead concentrations were found to have a higher propensity for delinquent and aggressive behavior by their teachers. Both pre- and postnatal blood lead levels in children proved remarkable predictors of delinquency in adolescents at age 20 and in adulthood. For every 5 µg increase in prenatal blood lead levels, there was a 40% increase in the risk of arrest [28].

New Media and the Risk of Violence Among Young People

Since violence is readily available in the media and the consumption of media violence is part of the everyday leisure repertoire of young males in particular, especially in the form of computer games, the question often arises whether the consumption of media violence also promotes aggressive and violent attitudes on the part of the consumers or can even make them addicted to such games. Media effects research presents a differentiated picture in this regard. For adults, playing violent video games showed no significant effects on later aggressiveness [29]. With regard to children and adolescents, however, the consensus is that media violence does contribute to real violent behavior, but that it is only one factor in a diverse bundle of causes. Only when combined with trauma or neglect in childhood, with a social environment conducive to violence, or with an incipient mental disorder does such media consumption can contribute to the development of violent crime [30]. According to estimates by criminologists and psychologists, about 5–10% of male adolescents represent a specific group that must be classified as particularly at risk, since here media representations of violence serve directly as patterns of identification and action [31, 32].

One particular risk group is adolescents with an a priori tendency toward aggressiveness, in whom pre-existing hostility toward others predisposes to the use of such games on PCs, smartphones or tablets. In particular, boys aged 11–17 with low levels of education found violent games particularly attractive and would identify with the game characters. There are also reports that the use of violent games by children without a corresponding pre-existing mentality leads to dissocial and more hostile attitudes in some of these children, and that violent media consumption fosters later aggression [32–34]. In addition to increased physical aggression, decreased empathy is often the result of such games [32, 35]. These studies suggest that intensive use of violent screen games in childhood and adolescence is a long-term risk factor for physical aggression, particularly affecting those adolescents who already tended to behave in this way before. Adolescents without such risk factors are hardly affected.

Predictability of Future Violence in Children and Adolescents?

In a small group of children, aggression with a tendency to violence can be observed as a persistent behavioral pattern [36]. In such children, genetic burden, a family environment favoring aggression, sometimes additional subtle brain developmental disorders and prenatal toxic influences often come together as a combination of several unfavorable causes.

A considerable number of studies conducted in the United States, Canada and Europe confirm that the combination of hereditary predisposition, birth complications, neurobiological risks, cognitive developmental deficits and desolate family environment results in a particularly unfavorable prognosis with regard to later dissocial and aggressive conspicuities [37–41].

Early functional features of the autonomic nervous system also appear to have prognostic significance. An unfavorable prognostic factor in children is an already low heart rate at the age of three. This has been explained by the fact that children with slower heart rates react less strongly to social restrictions, have less anxiety and fear fewer negative consequences of rule-breaking behavior. Some authors have therefore classified low heart rate as a biomarker for risk of future aggressivity [42]. Another study showed that lack of fear conditioning, as measured by the skin conductance response, at age three is a strong predictor of future criminal behavior. Three-year-old boys who did not show fear conditioning were significantly more likely to be conspicuous by criminal behavior at age 23 [43].

Children who have callous and unemotional traits (CU traits) show a particularly unfavorable developmental course [44]. These CU traits are very similar to the psychopathy traits that are also found in conscienceless adult perpetrators of violence. These include exaggerated dominance behavior, pathological lying, manipulative character traits, lack of guilt, lack of empathy, constant seeking of stimulation and early delinquency. The aggressiveness of such children and adolescents is more severe than that of children with social behavior disorder without such psychopathy criteria [44, 45].

In summary, in addition to social milieu and family background, the following biological risk factors for dissocial-aggressive behavior in childhood and adolescence have been identified:

- low heart rate [42]
- lack of fear conditioning [43]
- brain development disorders [41]
- prenatal malnutrition [24]
- prenatal nicotine and alcohol consumption by mother [22, 23]
- neuropsychological deficits, low intelligence [38]
- ADHD [19]
- hereditary disposition (family anamnesis) [17].

The previously held view that the earlier more multifaceted and more severe aggressive-dissocial behavior occurs, the higher the long-term risk of adolescent aggression and violence [46] must be relativized by more recent studies [47]. In a long-term analysis with questionnaires of adolescents on their own violent offenses, it was shown that the peak of such behavior is reached in the fourteenth year of life. At this age, 25% of male and 14% of female adolescents reported having committed violent offenses themselves in the course of a year (annual prevalence). Thereafter,

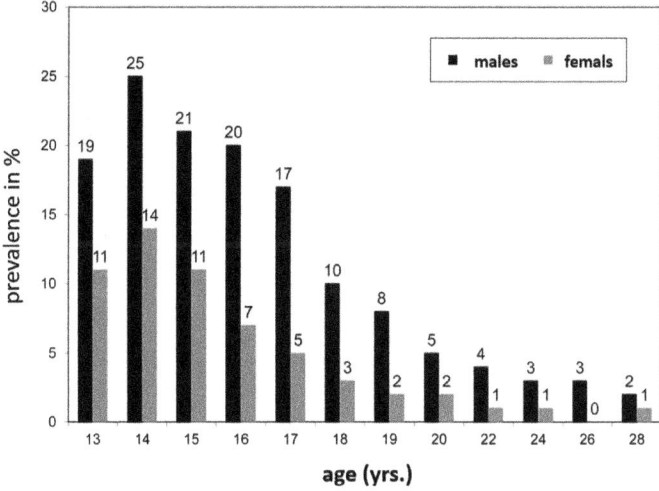

Fig. 15.1 Annual prevalence of self-reported violent offenses in youths and young adults. *Source* Boers [47].

there was a steady decline to 8% for male adolescents and 2% for female adolescents until the age of 18, and then a further decline to 2% and 1%, respectively, until the age of 28 (the end of the study period). The number of adolescents who had been classified as intensive offenders up to that point (approx. 6%) also decreased significantly in young adulthood [47] (see Fig. 15.1). The assumption that conspicuous delinquency in early childhood predicts later intensive delinquency is therefore only true to a limited extent, because not only the intensive offenders who later remain delinquent but also those who do not reoffend in adulthood are among the early conspicuous children.

Thus, a reliable prediction of long-term violent tendencies based on early childhood behavior is not possible in individual cases, even if there is a low statistical probability of higher delinquency.

References

1. Lohmeyer BA (2020) Youth and violent performativities; re-examining the connection between young people and violence, vol 1. Springer Nature Singapore Pte Ltd., Singapore
2. Knickmeyer RC, Gouttard S, Kang C et al (2008) A structural MRI study of human brain development from birth to 2 years. J Neurosci 28(47):12176–12182. https://doi.org/10.1523/JNEUROSCI.3479-08.2008
3. Odgers CL, Moffitt TE, Broadbent JM et al (2008) Female and male antisocial trajectories: from childhood origins to adult outcomes. Dev Psychopathol 20(02):673–716. https://doi.org/10.1017/S0954579408000333
4. Canino G, Polanczyk G, Bauermeister JJ, Rohde LA, Frick PJ (2010) Does the prevalence of CD and ODD vary across cultures? Soc Psychiatry Psychiatr Epidemiol 45(7):695–704.

https://doi.org/10.1007/s00127-010-0242-y
5. Ostrov JM, Keating CF (2004) Gender differences in preschool aggression during free play and structured interactions: an observational study. Soc Dev 13(2):255–277. https://doi.org/10.1111/j.1467-9507.2004.000266.x
6. Von Marées N, Petermann F (2010) Bullying in German primary schools. Sch Psychol Int 31(2):178–198. https://doi.org/10.1177/0143034309352416
7. UNESCO (2017) School Violenve and Bullying: Global Status Report. UNESCO. https://bangkok.unesco.org/content/school-violence-and-bullying-global-status-report-0
8. Campbell SB, Spieker S, Vandergrift N, Belsky J, Burchinal M (2010) Predictors and sequelae of trajectories of physical aggression in school-age boys and girls. Dev Psychopathol 22(01):133. https://doi.org/10.1017/S0954579409990319
9. Crick NR, Grotpeter JK, Bigbee MA (2002) Relationally and physically aggressive children's intent attributions and feelings of distress for relational and instrumental peer provocations. Child Dev 73(4):1134–1142. https://doi.org/10.1111/1467-8624.00462
10. Krischer MK, Sevecke K (2008) Early traumatization and psychopathy in female and male juvenile offenders. Int J Law Psychiatry 31(3):253–262. https://doi.org/10.1016/j.ijlp.2008.04.008
11. Theirworld.org. Explainer Child Soldiers. https://theirworld.org/explainers/child-soldiers. Published 2021
12. Unicef (2017) A familiar face. Violence in the lives of children and adolescents. https://data.unicef.org/wp-content/uploads/2017/10/EVAC-Booklet-FINAL-10_31_17-high-res.pdf
13. Odgers CL, Milne BJ, Caspi A, Crump R, Poulton R, Moffitt TE (2007) Predicting prognosis for the conduct-problem boy: can family history help? J Am Acad Child Adolesc Psychiatry 46(10):1240–1249. https://doi.org/10.1097/chi.0b013e31813c6c8d
14. Shalev I, Moffitt TE, Sugden K et al (2013) Exposure to violence during childhood is associated with telomere erosion from 5 to 10 years of age: a longitudinal study. Mol Psychiatry 18(5):576–581. https://doi.org/10.1038/mp.2012.32
15. Burt SA (2009) Are there meaningful etiological differences within antisocial behavior? Results of a meta-analysis. Clin Psychol Rev 29(2):163–178. https://doi.org/10.1016/j.cpr.2008.12.004
16. Rhee SH, Waldman ID (2002) Genetic and environmental influences on antisocial behavior: a meta-analysis of twin and adoption studies. Psychol Bull 128(3):490–529. http://www.ncbi.nlm.nih.gov/pubmed/12002699
17. Waldman ID, Tackett JL, Van Hulle CA et al (2011) Child and adolescent conduct disorder substantially shares genetic influences with three socioemotional dispositions. J Abnorm Psychol 120(1):57–70. https://doi.org/10.1037/a0021351
18. Teplin LA, Potthoff LM, Aaby DA, Welty LJ, Dulcan MK, Abram KM (2021) Prevalence, comorbidity, and continuity of psychiatric disorders in a 15-year longitudinal study of youths involved in the juvenile justice system. JAMA Pediatr e205807. https://doi.org/10.1001/jamapediatrics.2020.5807
19. Witthöft J, Koglin U, Petermann F (2010) Comorbidity of aggressive behavior and ADHD. Kindheit und Entwicklung. 19(4):218–227. https://doi.org/10.1026/0942-5403/a000029
20. Sobanski E (2006) Psychiatric comorbidity in adults with attention-deficit/hyperactivity disorder (ADHD). Eur Arch Psychiatry Clin Neurosci 256(S1):i26–i31. https://doi.org/10.1007/s00406-006-1004-4
21. van Lier PAC, Crijnen AAM (2005) Trajectories of peer-nominated aggression: risk status, predictors and outcomes. J Abnorm Child Psychol 33(1):99–112. https://doi.org/10.1007/s10802-005-0938-8
22. Toro R, Leonard G, Lerner JV et al (2008) Prenatal exposure to maternal cigarette smoking and the adolescent cerebral cortex. Neuropsychopharmacology 33(5):1019–1027. https://doi.org/10.1038/sj.npp.1301484
23. Fast DK, Conry J, Loock CA (1999) Identifying fetal alcohol syndrome among youth in the criminal justice system. J Dev Behav Pediatr 20(5):370–372. http://www.ncbi.nlm.nih.gov/pubmed/10533996

24. Neugebauer R, Hoek HW, Susser E (1999) Prenatal exposure to wartime famine and development of antisocial personality disorder in early adulthood. JAMA 282(5):455–462. http://www.ncbi.nlm.nih.gov/pubmed/10442661
25. Walsh WJ, Isaacson HR, Rehman F, Hall A (1997) Elevated blood copper/zinc ratios in assaultive young males. Physiol Behav 62(2):327–329. https://doi.org/10.1016/s0031-9384(97)88988-3
26. Dougherty DM, Moeller FG, Bjork JM, Marsh DM (1999) Plasma L-tryptophan depletion and aggression. Adv Exp Med Biol 467:57–65. https://doi.org/10.1007/978-1-4615-4709-9_7
27. Cherek DR, Lane SD, Pietras CJ, Steinberg JL (2002) Effects of chronic paroxetine administration on measures of aggressive and impulsive responses of adult males with a history of conduct disorder. Psychopharmacology 159(3):266–274. https://doi.org/10.1007/s002130100915
28. Wright JP, Dietrich KN, Ris MD et al (2008) Association of prenatal and childhood blood lead concentrations with criminal arrests in early adulthood. Balmes J (ed) PLoS Med 5(5):e101. https://doi.org/10.1371/journal.pmed.0050101
29. Kühn S, Kugler DT, Schmalen K, Weichenberger M, Witt C, Gallinat J (2019) Does playing violent video games cause aggression? A longitudinal intervention study. Mol Psychiatry 24(8):1220–1234. https://doi.org/10.1038/s41380-018-0031-7
30. Barlett C, Branch O, Rodeheffer C, Harris R (2009) How long do the short-term violent video game effects last?, vol 35. https://doi.org/10.1002/ab.20301
31. Bushman BJ, Huesmann LR (2006) Short-term and long-term effects of violent media on aggression in children and adults. Arch Pediatr Adolesc Med 160(4):348. https://doi.org/10.1001/archpedi.160.4.348
32. Anderson CA, Bushman BJ, Bartholow BD et al (2017) Screen violence and youth behavior. Pediatrics 140(Supplement 2):S142–S147. https://doi.org/10.1542/peds.2016-1758T
33. Gentile DA, Lynch PJ, Linder JR, Walsh DA (2004) The effects of violent video game habits on adolescent hostility, aggressive behaviors, and school performance. J Adolesc 27(1):5–22. https://doi.org/10.1016/j.adolescence.2003.10.002
34. Bender PK, Plante C, Gentile DA (2018) The effects of violent media content on aggression. Curr Opin Psychol 19:104–108. https://doi.org/10.1016/j.copsyc.2017.04.003
35. Möller I, Krahé B (2009) Exposure to violent video games and aggression in German adolescents: a longitudinal analysis. Aggress Behav 35(1):75–89. https://doi.org/10.1002/ab.20290
36. Wakschlag LS, Tolan PH, Leventhal BL (2010) Research review: 'Ain't misbehavin': towards a developmentally-specified nosology for preschool disruptive behavior. J Child Psychol Psychiatry 51(1):3–22. https://doi.org/10.1111/j.1469-7610.2009.02184.x
37. Kim-Cohen J, Caspi A, Taylor A et al (2006) MAOA, maltreatment and gene-environment interaction predicting children's mental health: new evidence and a meta-analysis. Mol Psychiatry 11(10):903–913. https://doi.org/10.1038/sj.mp.4001851
38. Raine A, Brennan P, Mednick S (1997) Interaction between birth complications and early maternal rejection in predisposing individuals to adult violence: specificity to serious, early-onset violence. Am J Psychiatry 154(9):1265–1271. https://doi.org/10.1176/ajp.154.9.1265
39. Bezdjian S, Raine A, Baker LA, Lynam DR (2011) Psychopathic personality in children: genetic and environmental contributions. Psychol Med 41(03):589–600. https://doi.org/10.1017/S0033291710000966
40. Raine A, Brennan P, Mednick B, Mednick SA (1996) High rates of violence, crime, academic problems, and behavioral problems in males with both early neuromotor deficits and unstable family environments. Arch Gen Psychiatry 53(6):544–549. http://www.ncbi.nlm.nih.gov/pubmed/8639038
41. Piquero A, Tibbetts S (1999) The impact of pre/perinatal disturbances and disadvantaged familial environments in predicting criminal offending, vol 8
42. Raine A (2015) Low resting heart rate as an unequivocal risk factor for both the perpetration of and exposure to violence. JAMA Psychiat 72(10):962. https://doi.org/10.1001/jamapsychiatry.2015.1364

43. Gao Y, Raine A, Venables PH, Dawson ME, Mednick SA (2010) Association of poor childhood fear conditioning and adult crime. Am J Psychiatry 167(1):56–60. https://doi.org/10.1176/appi.ajp.2009.09040499
44. Pardini DA, Lochman JE, Frick PJ (2003) Callous/unemotional traits and social-cognitive processes in adjudicated youths. J Am Acad Child Adolesc Psychiatry 42(3):364–371. https://doi.org/10.1097/00004583-200303000-00018
45. Frick PJ, Stickle TR, Dandreaux DM, Farrell JM, Kimonis ER (2005) Callous-unemotional traits in predicting the severity and stability of conduct problems and delinquency. J Abnorm Child Psychol 33(4):471–487. https://doi.org/10.1007/s10648-005-5728-9
46. Lahey B, Loeber R, Burke J, Rathouz P (2002) Adolescent outcomes of childhood conduct disorder among clinic-referred boys: predictors of improvement. J Abnorm Child Psychol 30:333–348
47. Boers K (2019) Delinquent behavior over the life course. Monatsschrift für Kriminologie und Strafrechtsreform 102(1):3–42. https://doi.org/10.1515/mks-2019-0004

Chapter 16
Rampage and School Shooting

Amok and terror are rare events, but they are all the more attention-grabbing because they are regularly accompanied by high numbers of victims. What is the probability of a rampage? Who are the perpetrators? Can such acts be prevented?

Difference Between Rampage and Terror

Rampages and spree killings are for outsiders completely incomprehensible, unpredictable, massive acts of violence, almost always carried out by male perpetrators, with the intention of killing as many as possible of selected or random victims. Unlike acts of terrorism, rampage killings are not motivated by political or ideological motives, but by personal ones. These killings often occur in a close social environment, whereas the targets of terrorist acts are symbolic of a system that is hated. Psychological disorders are common in both spree killers and lone terrorist offenders; the motives in both cases are often paranoid ideas. It is sometimes difficult to distinguish between rampage and terror in individual cases.

Deadliest Rampage Events of the Last 20 years

The world's most consequential rampages since 2000 with ten or more fatalities are listed below in chronological order, with a brief characterization of the perpetrators.

March 2001, Shijiazhuang, China
The 41-year-old perpetrator exploded four homemade bombs, killing 108 people and injuring 38 others. He had been deaf since the age of eight and was therefore a loner at school. The bombs detonated in front of the homes of his ex-wife and her family. In diary entries, he blamed them for his rape conviction and also for a 1994 traffic accident in which his mother was killed and his father injured [1].

June 2007, Baldayog City, Philippines
The 39-year-old perpetrator killed 10 people with a machete and injured 17 others. He was drunk during the rampage. In his village, he was considered an often drunk and mentally disturbed troublemaker who talked about being persecuted [2].

June/July 2007, Dnepropetrovsk, Ukraine
Two 19-year-old Ukrainians killed 21 mostly defenseless victims over a three-week period. This extraordinary series of murders began with the murder of a walker, whom they beat to death with a hammer. On the same day, they killed a man who was sleeping on a park bench. In the coming weeks, these "Dnepropetrovsk maniacs" committed 19 more murders. The two had tortured dogs and cats to death and filmed it even as minors. Since they also filmed the murders of humans, some of which appeared on the Internet, the motive was suspected to be the fun of killing and selling the murder videos on the Internet for money [3].

May 2012, Arqanqergen, Kazakhstan
The 19-year-old perpetrator shot 15 people at a border post on Kazakhstan's border with China. He then set fire to the building, requiring an extensive autopsy on the charred bodies. All but one of the 15 victims were work colleagues. He was the only person of Russian origin working at the border post, which is why experiences of discrimination were suspected as a motive[4].

July 2016, Kanagawa, Japan
The 26-year-old perpetrator, armed with knives, killed 19 residents and injured 26 others in a care facility for physically and mentally disabled people. In the run-up to the killing, he wrote a letter to the Speaker of the Japanese House of Representatives arguing that euthanasia should be allowed for the good of humanity. It would also strengthen the world economy and prevent World War III. He offered to kill disabled people, described his possible course of action and left his address. The authorities nevertheless classified him as harmless. He carried out his action as announced in the letter [5].

October 2017, Las Vegas, USA
This event is the deadliest mass murder by a single individual in the history of the United States. The 64-year-old perpetrator opened fire from the 32nd floor of a hotel on visitors to a music festival. He killed 60 people and 411 were injured. The perpetrator's body was found in the hotel by police along with 23 firearms he had taken there. According to investigators, he had lost a considerable amount of money while gambling in casinos. He was said to have suffered from depression and anxiety, but his doctor suspected he had bipolar disorder, for which he did not seek treatment [6].

November 2017, Sutherland Springs, Texas, USA
The 36-year-old perpetrator shot at members of the church community at the Sutherland Springs Church with a semi-automatic rifle for more than 11 min. He killed 27 people, including an unborn child, and injured 20 others. He had been dishonorably discharged from the Air Force after multiple offenses and repeatedly attracted attention for abuse and threats of violence against his female partners. He was a militant

atheist and specifically stormed that church to meet family members of his second wife who attended services there, after the couple had split up [7].

February 2020, Nakhon Ratchasima, Thailand
The 31-year-old perpetrator, a sergeant major in the Thai Army, shot and killed his superior and his mother-in-law at a military base because they had cheated him in a real-estate deal. He then forcibly looted the base's armory and stole a military jeep. In the following rampage, which he filmed and live-streamed on Facebook, he shot 30 people in a Buddhist temple and a shopping mall and injured 57 others [8].

February 2020, Hanau, Germany
The 43-year-old perpetrator shot 10 people from an immigrant background in front of a shisha bar. Subsequently, he shot his mother and himself. He had come to the attention of the authorities over a number of years with paranoid delusions and had been diagnosed with schizophrenia. This act is on the borderline between terror and rampage [9].

April 2020, Nova Scotia, Canada
The 51-year-old perpetrator killed 22 people and injured three others. After an argument with his partner as they left their anniversary party, he tied her up, returned to the party and shot seven guests. Subsequently, he drove through surrounding localities in a replica police car and shot people indiscriminately. He also set fire to several houses. He had already been convicted of assault in 2002 and had to undergo antiaggression training. Witnesses described him as paranoid, manipulative and jealous [10].

March 2021, Boulder, Colorado, USA
The 21-year-old perpetrator shot 10 people in a supermarket. His brother testified that he was prone to paranoid and antisocial behavior after being subjected to bullying in high school. He reportedly expressed to his wrestling team members that he was a victim of racism and Islamophobia [11].

Another particularly serious and unprecedented event that can be classified as an act of rampage—or rather a rampage flight—was the crash of the passenger plane Germanwings 9525 in the French Alps on March 24, 2015, deliberately caused by the 27-year-old German co-pilot. All passengers and the crew, 150 people in total, lost their lives. He was obviously suicidal and had been taking antidepressants at the time of the crash. Two weeks previously, a doctor had diagnosed him as possibly suffering from psychosis and recommended that he should be admitted to a psychiatric hospital [12].

The ages of the perpetrators of the events listed ranged from 19 to 64 years. All were male and most had a known psychiatric history. Triggering situations were often actual or perceived bullying or exclusion. However, bullying, social stress, mental disorder or generally normal psychologically comprehensible motives alone are not sufficient to explain such excessive acts of violence; tens of thousands of people worldwide have such experiences or mental problems without becoming mass murderers. There must be another brain disturbance that activates the violence center of the brain to a maximum through pathological mechanisms that still need to be explored in detail.

Most Momentous Rampage Killings Occur in the US

In the United States, such events are much more frequent than in other high-income countries. In recent years, between 269 (2014) and 610 (2020) mass shootings (four or more shot or killed in a single event, not including the shooter) have occurred each year in the US, with an upward trend from year to year [13]. There are also more mass shootings and many more killing spree fatalities in relation to the number of inhabitants in the US than in the rest of the world. One of several reasons for the significantly higher number of victims in the US is the easy access to firearms. About 80 percent of all people killed by firearms outside of armed conflicts were from the United States [14].

Deadliest School Shootings Worldwide

Probably the most dramatic form of rampage are school shootings involving young people using firearms at their own school. According to the School Shooter Info Database [15], there have been 147 such incidents worldwide in the last 50 years (since 1970). Almost all school shooters were male (nine were female). From decade to decade, there was a steady increase in such horror events until 2010: in the 1970s there were nine worldwide, this number rose to 20 in the 1980s, and in the 1990s there were 35.

April 1999, Littleton, USA
A school massacre with particular worldwide media impact was the Columbine High School shooting on April 20, 1999 in Littleton, USA. The two 18- and 17-year-old graduating seniors shot and killed 12 students aged 14 to 18, a teacher and finally themselves. Another 24 people were injured, some seriously. The perpetrators had planned an even greater act of bloodshed and planted an explosive device in the school by which hundreds of people were to be killed. However, as it did not explode due to a technical defect, they changed their plan at short notice and shot everyone they could find. The investigators assumed that the motives for the crime, which had been prepared for months, lay on the one hand in fantasies of grandiosity and the need to become famous, and on the other hand in revenge for actual or perceived bullying, lack of social acceptance and exclusion. The original videos of the crime, which can still be seen on the Internet, as well as the large number of victims, made this incident a particularly serious turning point, which on the one hand found a

number of imitators, and on the other hand triggered more intensive research into the causes of such acts [16].

After the massacre in 1999, the number of school shootings increased significantly due to copycat effects: by 2009, there were 42 worldwide, and from 2010 to 2020, another 45 cases occurred. Most of the perpetrators were below the age of 20 [15]. Their motivations varied from personal acts of revenge against people from their school or professional environment to the fun of killing. In contrast to individual terrorists, the preoccupation with ideologies does not play a role with school shooters.

The deadliest school massacres with more than 10 casualties after the Columbine High School event are listed below with brief descriptions of the respective perpetrators.

April 2002, Erfurt, Germany
The 19-year-old pupil shot 16 people and then himself at Gutenberg Grammar School. He had not been unpopular with his classmates, but he broke school rules by smoking and drinking, and skipped school by presenting fake medical certificates in order to train at the local shooting club instead. He was therefore expelled from school, but did not inform his parents. On the day of the crime, he appeared at the school with two pump guns and 500 rounds of ammunition, and shot 12 teachers, a staff member, a policeman and two fellow students. He probably wanted to take revenge for having been expelled from school without qualifications. He had previously said that he wanted to become as famous as the Columbine High School shooters [17].

January 2007, Blacksburg, Virginia, USA
In the deadliest school shooting ever in the US, the 23-year-old perpetrator shot 33 people, including himself, and injured 17 others at Virginia Polytechnic Institute. He had previously been diagnosed with anxiety disorder, selective mutism and depression, for which he was receiving treatment [18].

September 2008, Kauhajoki, Finland
The 22-year-old perpetrator shot 11 people (including himself) and injured as many at Seinäjoki University, where he was a student. He was the victim of bullying at school and suffered from anxiety disorders, panic attacks and depression. He was said to have shown symptoms of anxious-avoidant personality disorder and schizotypal personality disorder [19].

March 2009, Winnenden, Germany
The 17-year-old student shot 15 people at a secondary school, and 11 were seriously injured. He was described by fellow pupils as quiet and withdrawn and was undergoing treatment by a youth psychiatrist for depression. More than 200 pornographic pictures were found on his computer, as well as first-person shooter games such as Counter-Strike, in which as many opponents as possible have to be killed by using firearms. When the police arrived, he fled from the school. During his escape, he shot a workman in the nearby psychiatric center. He then took the driver of a car hostage and fled with him at gunpoint. The hostage reported that he said to him, "I have already killed 15 people in my old school and that was not all for today." Later he is said to have asked "Do you think we can find another school?" When asked

why he was doing all this, he replied, "For fun, because it is fun!" Finally he shot himself [20].

December 2012, Newtown, Connecticut, USA
The 20-year-old perpetrator shot 26 people at a primary school, including 20 children between the ages of six and seven, and finally himself. He had attended this primary school as a child and was diagnosed with Asperger's syndrome, depression, anxiety and obsessive–compulsive disorders. He had access to various firearms through his mother, whom he also shot. Why he chose the primary school for his crime remained unexplained [21].

February 2018, Parkland, Florida, USA
The 19-year-old perpetrator shot 17 people at his former high school. Prior to the crime, he changed schools six times in three years, including a school for children with emotional and learning disabilities, repeatedly attracting attention with threats of violence. Consultants diagnosed him with depression, autism and ADHD. However, he did not receive in-patient treatment [22].

May 2018, Santa Fe, Texas, USA
The 17-year-old perpetrator shot 10 people and injured 14 others at a high school in Texas. He was described by fellow students as friendly and funny. On Facebook, however, he staged himself as being close to violence, for example wearing a shirt with the motto "Born to Kill." His father suspected that bullying was the reason for the crime [23].

October 2018, Kerch, Crimean Peninsula, Ukraine
The 18-year-old perpetrator stormed Polytech College with a shotgun and 150 rounds of ammunition. He shot wildly, targeting computer monitors and fire extinguishers as well as people, in addition to detonating a large nail bomb. In total, he killed 20 people and injured 70 others. He was a loner whose father was described as extremely aggressive. The mother isolated her son through various prohibitions. On social media, he had contact with groups themed around mass murderers, including a fan club of the Columbine school shooting [24].

March 2019, Suzano/Sao Paulo, Brazil
Two perpetrators, aged 17 and 25, opened fire in their former school, killing six students, two school employees and then themselves. Before that, they had shot the uncle of the 17-year-old perpetrator. Both had previously been bullied by fellow pupils. The crime was planned for a year and was inspired by the Columbine High School massacre [25].

The events depicted here are only the most consequential school shootings. Worldwide, 45 school shootings have been reported since 2010.

Who Becomes a Mass Murderer?

To clarify the question of why such acts occur at all, a number of studies have been conducted on the psychosocial situation, psychopathology and motivational starting point of the perpetrators [26–28].

Studies of Surviving Spree Killers

About half of the spree killers shoot themselves or are shot at the end of the killing. Obtaining reliable information about their psychological situation before the crime is difficult, if not impossible in many such cases. A comprehensive analysis of the psychological and social situation of surviving spree killers in Germany was carried out for a 30-year period (1980–2010) [26], during which there were 123 rampage killings; court records of 44 surviving rampage killers could be obtained. Unlike those who killed themselves or were shot, these perpetrators were investigated by public prosecutors. Therefore, not only press reports but also extensive court files including psychiatric expertise were available, from which the initial psychological profile before the crime could be reconstructed.

A review of the files revealed that about 80 percent of the killers had a previous psychiatric illness, such as schizophrenic psychoses, alcohol or drug abuse, narcissistic and paranoid personality disorders, but also depressive illnesses, often combined with suicidal tendencies.

From the file analysis of the 44 cases, three types of perpetrators were crystallized [26].

1. Fourteen adolescents or young adults who wanted to take revenge on teachers and classmates out of anger and despair because of the humiliations and exclusions they felt, often after a personal failure (not being promoted to a higher class, changing schools). Almost all of them were diagnosed with a mental disorder in the form of depression, anxiety disorder or ADHD, as well as paranoid and narcissistic personality traits.
2. Sixteen psychotic persons (average age 30 years) with schizophrenia, in several cases combined with alcoholism. These perpetrators wanted to defend themselves against alleged attackers or take revenge due to delusions of threat or persecution. Some were under the influence of religious delusions when committing the crime.
3. Fourteen middle-aged men (average age 47 years) with narcissistic, paranoid and dissocial personality traits, half of whom were under the influence of alcohol during the crime. Motives for the crime were anger, hatred and revenge after perceived slights due to prolonged conflicts within the family or at work.

About one-third of the perpetrators (second group) of this analysis suffered from a diagnosed schizophrenic psychosis. This corresponds to the proportion of psychotic offenders reported by other research groups [27].

Further Research Projects on the Psychology of Rampage Killers

In a project funded by the German Ministry of Education and Research for the criminological analysis of rampage crimes (TARGET project), 19 cases of young rampage killers and 39 cases of adult rampage killers were studied for the period March 2013 to June 2016 [27, 29]. The age range for the young rampage killers was between 14 and 23 years. Eight of these offenders killed themselves following the crime; four attempted to kill themselves. In 10 of these 19 acts, where the perpetrators used firearms, there were 44 deaths. In the nine cases in which the perpetrators used other means (e.g. cutting and stabbing weapons), there were six fatalities.

Among the core group of young spree killers, the majority were quiet and withdrawn personality types who quickly felt slighted and disregarded. In contrast to other violent offenders in their age group, the later spree killers were not impulsive, aggressive or dissocial. They did not show the typical accumulation of risk characteristics usually present in violent, aggressive youths. Most of them did not stand out in school and among peers due to social behavior disorders, violence or aggression. The social background was mostly unremarkable, the families tended to belong to the middle class and were often financially well off, there were no "broken homes" with violence and social neglect; the parents were usually concerned about the welfare of their children. During puberty, there was increased social withdrawal; most of the later perpetrators were considered shy, quiet, fearful and unsociable, but on the other hand, they developed an excessive interest in violent films and computer games as well as martial music, former assassinations and rampages. In addition, there were expressions of excessive need for revenge and hatred, for example, in their diaries. They identified with earlier perpetrators and staged their acts as revenge for subjectively experienced slights. The acts were usually preceded by long-term planning. Over the course of months, they developed pronounced fantasies of violence and killing.

Often there were warning signs of problematic personality development, with hints of the planned act to peers.

In summary, the report states that thoughts of rage, hatred and generalized revenge cannot be explained in terms of normal psychology, but only from a psychopathological perspective. The typical killing motives of dissocial criminal murderers, such as greed, power over the victim, or sadistic and sexually abnormal fantasies, as well as the intention to escape prosecution after the crime, did not play a role in young spree killers [27].

The same research project examined 40 adult spree killers, including two women. Of these 40 offenders, 14 were paranoid-schizophrenic at the time of the crime. In 15 cases, a personality disorder of the paranoid and narcissistic type was diagnosed. Typical characteristics of a paranoid personality disorder are misinterpretations of experiences with a hostile attitude and quarrelsome behavior [27]. Narcissistic personality disorders are characterized by a grandiose sense of one's own importance, a desire for excessive admiration, a lack of empathy and a deep-rooted sensitivity to offenses (see Chap. 10).

What young and adult amok perpetrators had in common was that almost all of them were men or male adolescents with a massive psychological fragility, paranoid attitudes, fantasies of violence and killing, and an affinity for weapons.

School Shooting Preventive Measures

After such rampages, the question is regularly raised as to whether potential perpetrators might not have shown some warning signs or constellations of symptoms at an early stage, which would make the acts preventable. Since psychotic developments can also occur in young people and adults who have been largely inconspicuous and psychologically healthy up to that point, and since the perpetrators also plan their acts silently, it is very difficult to recognize potential spree killers in advance. However, efforts to develop risk profiles for pupils and young people seem to be promising, with the aim of providing them in time with counseling, psychological care or, if necessary, medical specialist treatment.

Another working group from the TARGET project addressed the question of what criteria can be identified for the early detection of serious targeted violence in schools [29, 30]. Based on the realization that serious acts of violence in schools are always the endpoints of a crisis-like development of the perpetrators, usually lasting for years, which must be recognized in advance in order to implement effective prevention measures, it seemed important that the perpetrator's environment (parents, peers, teachers) perceive crisis symptoms or delusional behavior and react to them. In this context, it was also important to find decision criteria to distinguish substantial (potentially serious) from fleeting threats and announcements. The authors analyzed 11 cases of serious targeted violence in schools for the period 1999–2013. In six of the 11 cases studied, the course of the offense was terminated by suicide. During the same period, threats of rampage were made by 115 pupils. Among them were 10 substantial threats in which there were concrete preparations for the crime which carried a considerable risk of being found out. In the 105 other cases, there were fleeting threats with no verifiable preparations to commit a crime and no serious risk. In a first comparison with pupils who only made fleeting threats and had not yet committed a crime, the perpetrators of serious acts of school violence showed significantly more planning behavior, were more fixated on an injustice and more likely to identify with previous perpetrators of violence, and more often made announcements in the run-up to the execution of the act.

In the 11 main cases of severe targeted school violence investigated, the perpetrators experienced a severe crisis early on, which strongly influenced their further development. The timespan between the onset of the crisis and the execution of the offense was on average five years. Evidence was found that threats and crime announcements made in the peer context were often not passed on to school staff [31].

The authors recommended the introduction of structured interview guidelines with which both the severity, content and duration of violent fantasies as well as the seriousness of planned actions can be ascertained and diagnostically classified, if necessary with the help of child and youth psychiatric expertise. Only the continuous cooperation of schools, counseling centers, youth welfare offices, specialized clinics and therapists in private practice can help to avert violent, crisis-like developments [31]. To realize this, the NETWASS prevention project for the training of teachers and other educational staff was established, with a center in Berlin [31], as well as The Amok Prevention Counselling Network at the University of Giessen [32].

Early Warning Symptoms: "Leaking"

School shooters no longer manage to gain recognition through sporting or school achievements or successes within the peer group, which is why age-appropriate strategies cannot be used to gain self-esteem. Instead, they see self-dramatization and the spectacular execution of extreme violence as a proven means to this end. A variety of risk factors for serious school violence, usually occurring simultaneously, have been elaborated so far [31]. These include bullying/social exclusion at school, permanent conflicts with teachers, suicidal tendencies, social withdrawal, drop in performance at school, openly aggressive behavior, conspicuous change in behavior, communicated fantasies of revenge and violence, interest in weapons, fascination with school shootings, excessive consumption of violent media, and especially also "leaking." This means that potential school shooters often make announcements or drop hints about the planned offense to classmates, acquaintances or even educators. These clues resulting from leaking are the most important indicator for early detection of conspicuous youths and cause for early intervention.

However, the number of threats that were carried out is much higher than the number of announcements that were not linked to a concrete plan to commit the crime. In a Berlin study, 427 threats of lethal violence were registered in schools over a eleven-year period (1996–2007) [33]. This corresponded to a rate of 0.3 threats per 1000 pupils in a school year. Ninety percent were male pupils; 18 of these referred to the 2002 school shooting in Erfurt. All cases were dealt with within the school committees by teachers, social workers, psychologists or with the involvement of external specialists.

It is obvious that the evidence-based counseling and training networks for the prevention of serious violence in schools [31] based on these findings contributed to the fact that in the last ten years, in contrast to the previous decade in Germany and to the conditions in the US, there were no more serious rampages in schools (status April 2021).

Warning Symptoms in Adult Spree Killers

Another study of 33 adult spree killers (94% male) for the period 2000–2012 showed that almost half had no regular job, but had financial problems and a criminal record [34]. The majority were psychologically traumatized due to loneliness in early childhood. Half planned the crime weeks, months or even years in advance. A high proportion were under the influence of alcohol or drugs during the course of the attack. One-third suffered from a schizophrenic psychosis. In all cases, one or more warning signs in the form of "leaking," personality change with an increase in aggressive behavior, threats to the subsequent victims, preoccupation with violent scenarios and interest in weapons could be determined retrospectively for the time before the crime. The acts therefore occurred because these omens were not classified by anyone as warning signs of a real danger.

A US study of all mass shootings from 2009 to 2019 found [35] that almost half of the perpetrators were suicidal; 40 percent committed suicide as the final act of the rampage; 10 percent were shot by security forces. More than half of the perpetrators hinted at their intentions to those close to them before committing the offense, so there were warning signals there too that were not sufficiently heeded. If they had been taken seriously, the authors estimate that half the number of fatalities could have been prevented [35].

Which Brain Functions Are Damaged in Spree Killers?

The risk constellations described for rampages and school shootings, such as actual or perceived insults, traumatizing life events, feelings of exclusion, conflicts at school, in the family or at the workplace, suicidal thoughts, social withdrawal, loss of performance, conspicuous changes in behavior, fantasies of revenge and violence, excessive consumption of violent media and even psychotic developments are present in many people without them committing extreme acts of violence. If one looks at rampage events in relation to the total number of men in the age group considered for this, then, depending on the author and the time period studied, one person in 5–10 million becomes a rampage offender [34], in other words less than 0.000001 percent of the relevant male age range. These are therefore extremely exceptional cases, which is why the question arises what distinguishes these people from those who have similar risk characteristics but do not become extreme violent offenders.

Often, attempts at explanations are made that are within the realm of psychological comprehensibility, such as the murder of narcissistic and paranoid traits, revenge for slights or fantasies of violence. However, such emotions occur in thousands of adolescents and adults without them becoming spree killers. The extreme rarity of such events suggests that common, normal psychological causes are not sufficient to explain them. There must be a rare constellation of psychopathological phenomena that have to be traced back to pathological changes in certain brain functions that occur only in very rare cases.

So far, neuropathological examinations have been performed on only two brains of spree killers. These two cases are also among the most historically significant (see Chap. 7). The first is the schoolteacher Ernst Wagner, who killed his wife and four children in 1913 near Stuttgart under the influence of delusional thoughts, then shot nine people and injured 11. Wagner spent the years after the crime until his death as a forensic psychiatric patient in a clinic in South Germany.

The second case is Charles Whitman, who shot 17 people in Texas in 1966, injured another 32, and then was himself shot by the police (see Chap. 7). In both mass murderers, brain pathological examinations revealed damage to a site of the medial temporal brain directly adjacent to the amygdala (see Fig. 7.3). This brain region is of central importance for the control of the amygdala and the brain aggression centers in the hypothalamus, which can become unrestrainedly active if the control by the areas of the temporal and frontal brain responsible for this are no longer functioning. Similar defects of the brain substance have been demonstrated by magnetic resonance imaging and computed tomography studies in murderers and other serious violent offenders [36–42].

It is probable that the amygdala and the brain centers of aggression located in the hypothalamus (see Chap. 6, Figs. 6.4, 6.6 and 6.7) not only become active in response to normal psychologically comprehensible provocations, but can also develop pathological overactivity as a result of a primary disease of these brain areas themselves or of upstream inhibitory brain regions.

In psychiatry, there are many examples of psychosyndromes that represent pathological exaggerations of emotions that are quite reasonable in a healthy state. These include anxiety, euphoria or even disgruntlement, anger, rage and mistrust, as well as situation-appropriate aggression. Of all these essential feelings for coping with life there are pathological aberrations like panic attacks and phobia, mania, depression, paranoid psychoses and also violent excesses. The brain-biological causes of these psychosyndromes, which are also suspected for extreme violence in the form of rampage, are still partly unknown. However, there is increasing evidence that inflammation-like processes similar to those present in autoimmune diseases (attacks by the immune system on the body's own organs), such as in multiple sclerosis, can damage the brain areas relevant for these emotions. Such neuroimmunological causes attacking the brain tissue have already been demonstrated in depressive, manic and paranoid-psychotic patients [43–46] and are also suspected in pathological forms of aggression. It is therefore probable that in rare cases the nerve cell groups relevant to aggression in the hypothalamus (belonging to the reptilian part of our brain) become spontaneously active even without external cause due to various disease processes, or that the threshold for aggressive discharge is thereby lowered to such an extent that even minimal triggers are sufficient to provoke excessive violent action.

One-third of adult spree killers have been diagnosed with schizophrenic psychoses. Here, the affected persons may be prompted to act by hallucinations or commanding voices, or they take revenge for delusionally perceived threats, persecution and ridicule. The increased tendency to violence of some schizophrenics can be explained on the one hand by these delusional phenomena and on the other hand by the fact that the brain regions which were also damaged in Wagner, Whitman and other murderers and which are of central importance for the inhibition of the aggression centers are also affected by the disease [39, 47].

Rampages in the Prodromal Stages of Schizophrenic Disorders

Schizophrenic illnesses do not occur suddenly like epileptic seizures, heart attacks or strokes, but have long subtle preceding stages (so-called prodromal stages) which last for months and even years and often begin in adolescence after puberty. In the prodromal stage, the typical symptoms of the disease, such as hallucinations, delusions, paranoid ideas of threats, persecution and disorganized thinking, do not yet appear, but develop very slowly until the typical psychopathological appearance in young adulthood, and more rarely after a gradual progression, even in adulthood. Due to the lack of typical symptoms, it is difficult even for psychiatrists to correctly diagnose such prodromal phases of schizophrenic disorders; in most cases, they are not recognized as such at all. The affected persons, until then inconspicuous adolescents, withdraw, experience a wobble in their life line, often tend to depressiveness, become more suspicious, sensitive and irritable, or have a feeling of being excluded by others, without this corresponding to the real circumstances. They go into strange worlds of thought which can manifest as preoccupation with occult things or may in some cases develop mania-like fantasies of grandeur, often connected with a drop in performance at school. These prodromal symptoms match the behavioral profile described in the run-up to the rampage in some school shooters and suggest that an unrecognized prodromal stage of schizophrenic psychosis was present.

There are numerous reports, especially in the older psychiatric literature, of homicides, including rampages and acts of terrorism (then called "political murders"), in the prodromal stage of schizophrenia [48]. Some perpetrators felt an obsessive drive to murder others, believed they had to act on higher inspiration and felt relieved after the act. Recent studies conducted in the United States also show that homicidal acts in the run-up to schizophrenic illness are as common as in clinically fully developed psychoses [49]. Even if young offenders have not yet been diagnosed with schizophrenic psychosis, the presence of a prodromal stage is likely if any of the behaviors described are present.

Hate and pathological aggressiveness need not be the result of psychotic or prepsychotic development or even brain tissue damage in every case. Massive and prolonged psychological trauma, especially in early childhood, can also lead to lifelong dysfunction of brain areas that inhibit violent behavior. In some cases, both early traumatization and psychosis coincide. If acutely stressful situations are then added, the risk of engaging in extreme violence is particularly high.

Future Risk of Rampages and School Shootings

If parents, teachers and educators are consistently made aware of the early warning signs of an emerging change in a young person's character with a potential tendency toward violence, and if the appropriate counseling networks are consulted in good time [31, 32], there is hope that the incidence of school shootings will decrease.

The psychosocial and psychopathological situation and the resulting risk constellations of adult spree killers are similar to those of lone-actor terrorists (see Chap. 17). In recent years, internationally tested risk assessment scales have been developed for the early detection and prevention of acts committed by adults with a tendency toward extreme violence [50–52]. The extent to which these can contribute to curbing the phenomenon of mass murder remains to be seen.

It is to be feared that some perpetrators will continue to hatch their plans in silence, unnoticed by the outside world, and that high numbers of victims will shock bystanders, survivors and the public.

References

1. Wikipedia.org. Shijiazhuang bombings. Wikipedia.org. https://en.wikipedia.org/wiki/Shijiazhuang_bombings. Published 2021. Accessed 1 Apr 2021
2. Wikipedia.org. Danilo Guades. Wikipedia.org. https://en.wikipedia.org/wiki/Danilo_Guades. Published 2020. Accessed 12 Feb 2021
3. Wikipedia.org. Dnepropetrovsk maniacs. Wikipedia.org. https://en.wikipedia.org/wiki/Dnepropetrovsk_maniacs. Published 2021. Accessed 1 Apr 2021
4. Wikipedia.org. Arkankergen mass murder. Wikipedia.org. https://en.wikipedia.org/wiki/Arkankergen_mass_murder. Published 2021. Accessed 1 Apr 2021
5. Wikipedia.org. Sagamihara stabbings. Wikipedia.org. https://en.wikipedia.org/wiki/Sagamihara_stabbings. Published 2021. Accessed 1 Apr 2021
6. Wikipedia.org. 2017 Las Vegas shooting. Wikipedia.org. https://en.wikipedia.org/wiki/2017_Las_Vegas_shooting. Published 2021. Accessed 1 Apr 2021
7. Wikipedia.org. Sutherland Springs church shooting. Wikipedia.org. https://en.wikipedia.org/wiki/Sutherland_Springs_church_shooting. Published 2021. Accessed 1 Apr 2021
8. Wikipedia.org. Nakhon Ratchasima shootings. Wikipedia.org. https://en.wikipedia.org/wiki/Nakhon_Ratchasima_shootings. Published 2021. Accessed 1 Apr 2021
9. Wikipedia.org. Hanau shootings. https://en.wikipedia.org/wiki/Hanau_shootings. Published 2021
10. Wikipedia.org. 2020 Nova Scotia attacks. Wikipedia.org. https://en.wikipedia.org/wiki/2020_Nova_Scotia_attacks. Published 2021. Accessed 1 Apr 2021
11. Wikipedia.org. 2021 Boulder shooting. Wikipedia.org. https://en.wikipedia.org/wiki/2021_Boulder_shooting. Published 2021. Accessed 1 Apr 2021
12. Wikipedia.org. Germanwings flight 9525. https://en.wikipedia.org/wiki/Germanwings_Flight_9525. Published 2021
13. Gunviolencearchive.org. Past summary ledgers. gunviolencearchive.org. https://www.gunviolencearchive.org/past-tolls. Published 2021. Accessed 28 Apr 2021
14. Grinshteyn E, Hemenway D (2016) Violent death rates: The US compared with other high-income OECD Countries, 2010. Am J Med 129(3):266–273. https://doi.org/10.1016/j.amjmed.2015.10.025

15. Langman P (2020) schoolshooters.info. https://schoolshooters.info/search-database. Published 2020. Accessed 13 Mar 2020
16. Wikipedia.org. Columbine High School massacre. Wikipedia.org. https://en.wikipedia.org/wiki/Columbine_High_School_massacre. Published 2021. Accessed 28 Apr 2021
17. Wikipedia.org. Erfurt school massacre. Wikipedia.org. https://en.wikipedia.org/wiki/Erfurt_school_massacre. Published 2021. Accessed 28 Apr 2021
18. Wikipedia.org. Seung-Hui Cho. Wikipedia.org. https://en.wikipedia.org/wiki/Seung-Hui_Cho. Published 2021. Accessed April 1, 2021.
19. Wikipedia.org. Kauhajoki school shooting. Wikipedia.org. https://en.wikipedia.org/wiki/Kauhajoki_school_shooting. Published 2021. Accessed 1 Apr 2021
20. Wikipedia.org. Winnenden school shooting. Wikipedia.org. https://en.wikipedia.org/wiki/Winnenden_school_shooting. Published 2021. Accessed 28 Apr 2021
21. Wikipedia.org. Sandy Hook Elementary School shooting. Wikipedia.org. https://en.wikipedia.org/wiki/Sandy_Hook_Elementary_School_shooting. Published 2021. Accessed April 1 Apr 2021
22. Wikipedia.org. Stoneman Douglas High School shooting. Wikipedia.org. https://en.wikipedia.org/wiki/Stoneman_Douglas_High_School_shooting#Suspect. Published 2021. Accessed 1 Apr 2021
23. Wikipedia.org. Santa Fe High School shooting. Wikipedia.org. https://en.wikipedia.org/wiki/Santa_Fe_High_School_shooting. Published 2021. Accessed 1 Apr 2021
24. Wikipedia.org. Kerch Polytechnic College massacre. Wikipedia.org. https://en.wikipedia.org/wiki/Kerch_Polytechnic_College_massacre. Published 2021. Accessed 12 Feb 2021
25. Wikipedia.org. Suzano school schooting. Wikipedia.org. https://en.wikipedia.org/wiki/Suzano_school_shooting. Published 2021. Accessed 12 Feb 2021
26. Peter E, Seidenbecher S, Bogerts B, Dobrowolny H, Schöne M (2019) Mass murders in Germany—classification of surviving offenders based on the examination of court files. J Forens Psychiatry Psychol. 30(3):381–400. https://doi.org/10.1080/14789949.2019.1593486
27. Bannenberg B. (2017) Schlussbericht Projekt TARGET. Teilprojekt Gießen: Kriminologische Aspekte von Amoktaten—junge und erwachsene Täter von Amoktaten, Amokdrohungen (Final report project TARGET. Subproject Giessen: criminological aspects of amok deeds—young and adult perpetrators). Bundesministerium für Bildung und Forschung. https://www.uni-giessen.de/fbz/fb01/professuren-forschung/professuren/bannenberg/news/Schlussbericht-Target-Giessen. Published 2017. Accessed 3 May 2021
28. Böckler N, Leuschner V, Zick A, Scheithauer H (2018) Same but different? Developmental pathways to demonstrative targeted attacks-qualitative case analyses of adolescent and young adult perpetrators of targeted school attacks and Jihadi terrorist attacks in Germany. Int J Dev Sci 12(1–2):5–24. https://doi.org/10.3233/DEV-180255
29. Leuschner V, Fiedler N, Schultze M et al (2017) Prevention of targeted school violence by responding to students' psychosocial crises: the NETWASS program. Child Dev 88(1):68–82. https://doi.org/10.1111/cdev.12690
30. Leuschner V, Bondü R, Schroer-Hippel M et al (2011) Prevention of homicidal violence in schools in Germany: The Berlin leaking project and the Networks Against School Shootings Project (NETWASS). New Dir Youth Dev 2011(129):61–78. https://doi.org/10.1002/yd.387
31. Fiedler N, Sommer F, Leuschner V, Scheithauer H (2019) Student crisis prevention in schools: the NETWorks Against School Schootings Programm (NETWASS)—an approach suitable for the prevention of violent extremism? Int J Dev Sci 13:109–122. https://www.researchgate.net/publication/338808615_Student_Crisis_Prevention_in_Schools_The_NETWorks_Against_School_Shootings_Program_NETWASS_-_An_Approach_Suitable_for_the_Prevention_of_Violent_Extremism
32. Bannenberg B (2017) Beratungsnetzwerk Amokprävention—Ein wissenschaftsbasiertes Beratungsangebot zur Amokprävention (Amok prevention consulting network—a science-based consulting service for amok prevention). uni-giessen.de. https://www.uni-giessen.de/fbz/fb01/professuren-forschung/professuren/bannenberg/mediathek/dateien/beratungsnetzwerk-amok-2016.pdf. Published 2017. Accessed 3 May 2021

33. Leuschner V, Bondü R, Allroggen M, Scheithauer H (2016) Leaking: Frequency and correlates of announcements and threats of homicidal violence reported by Berlin schools between 1996 and 2007. Z Kinder Jugendpsychiatr Psychother 44(3):208–219. https://doi.org/10.1024/1422-4917/a000423
34. Allwinn M, Hoffmann J, Meloy JR (2019) German mass murderers and their proximal warning behaviors. J Threat Assess Manag 6(1):1–22. https://doi.org/10.1037/tam0000122
35. Everytownresearch.org. Ten Years of Mass Shootings in the United States, An Everytown for Gun Safety Support Fund Analysis. https://everytownresearch.org/massshootingsreports/mass-shootings-in-america-2009-2019/. Published 2019. Accessed 23 Mar 2020
36. Schiltz K, Witzel JG, Bausch-Hölterhoff J, Bogerts B (2013) High prevalence of brain pathology in violent prisoners: a qualitative CT and MRI scan study. Eur Arch Psychiatry Clin Neurosci 263(7):607–616. https://doi.org/10.1007/s00406-013-0403-6
37. Witzel JG, Bogerts B, Schiltz K (2016) Increased frequency of brain pathology in inmates of a high-security forensic institution: a qualitative CT and MRI scan study. Eur Arch Psychiatry Clin Neurosci 266(6):533–541. https://doi.org/10.1007/s00406-015-0620-2
38. Cope LM, Ermer E, Nyalakanti PK, Calhoun VD, Kiehl KA (2014) Paralimbic gray matter reductions in incarcerated adolescent females with psychopathic traits. J Abnorm Child Psychol 42(4):659–668. https://doi.org/10.1007/s10802-013-9810-4
39. Anderson NE, Kiehl KA (2013) Psychopathy and aggression: when Paralimbic dysfunction leads to violence. In: Neuroscience of aggression, pp 369–393. https://doi.org/10.1007/7854_2013_257
40. Raine A, Buchsbaum M, Lacasse L (1997) Brain abnormalities in murderers indicated by positron emission tomography. Biol Psychiatry 42(6):495–508. https://doi.org/10.1016/S0006-3223(96)00362-9
41. Raine A (2013) The anatomy of violence—the biological roots of crime. Penguin Books, London
42. Yang Y, Glenn AL, Raine A (2008) Brain abnormalities in antisocial individuals: implications for the law. Behav Sci Law 26(1):65–83. https://doi.org/10.1002/bsl.788
43. Goldsmith DR, Rapaport MH, Miller BJ (2016) A meta-analysis of blood cytokine network alterations in psychiatric patients: comparisons between schizophrenia, bipolar disorder and depression. Mol Psychiatry 21(12):1696–1709. https://doi.org/10.1038/mp.2016.3
44. Al-Diwani AAJ, Pollak TA, Irani SR, Lennox BR (2017) Psychosis: an autoimmune disease? Immunology 152(3):388–401. https://doi.org/10.1111/imm.12795
45. Bogerts B, Winopal D, Schwarz S et al (2017) Evidence of neuroinflammation in subgroups of schizophrenia and mood disorder patients: a semiquantitative postmortem study of CD3 and CD20 immunoreactive lymphocytes in several brain regions. Neurol Psychiatry Brain Res 23:2–9. https://doi.org/10.1016/j.npbr.2016.11.001
46. Miller BJ, Goldsmith DR (2017) Towards an immunophenotype of schizophrenia: progress, potential mechanisms, and future directions. Neuropsychopharmacology 42(1):299–317. https://doi.org/10.1038/npp.2016.211
47. Bogerts B, Schöne M, Breitschuh S (2018) Brain alterations potentially associated with aggression and terrorism. CNS Spectr 23(2):129–140. https://doi.org/10.1017/S1092852917000463
48. Nielssen O, Large M, Ryan C, Hayes R (2007) Legal implications of the increased risk of homicide and serious violence in the first episode of psychotic illness. 31 Crim LJ 287. https://www.publicdefenders.nsw.gov.au/Documents/legalimplications.pdf. Published 2007. Accessed 3 May 2021
49. Mojtabai R (2006) Psychotic-like experiences and interpersonal violence in the general population. Soc Psychiatry Psychiatr Epidemiol 41(3):183–190. https://doi.org/10.1007/s00127-005-0020-4
50. Koller S (2020) Good practices in risk assessment for terrorist offenders. German Council on Foreign Relations. https://dgap.org/en/research/publications/good-practices-risk-assessment-terrorist-offenders. Published 2020.

51. Rma.scot. Violent Extremist Risk Assessment 2 Revised (VERA 2R). Risk management authority2019. https://www.rma.scot/wp-content/uploads/2019/09/RATED_VERA-2_July-2019_Hyperlink-Version.pdf. Published 2019.
52. Meloy JR, Gill P (2016) The lone-actor terrorist and the TRAP-18. J Threat Assess Manag 3(1):37–52. https://doi.org/10.1037/tam0000061

Chapter 17
Terror

What is Terror?

Almost weekly we receive news about terrorist attacks and everyone has an idea of what terror is. Nevertheless, there is no consensus worldwide on what can be called terror. Depending on the political, ideological, religious and social system, opinions differ. Accordingly, the term is used vaguely and variably in different political contexts. Terrorists are not infrequently celebrated as freedom or resistance fighters or even as heroes by the adherents of the ideology in whose name they commit acts of terror, and may even be revered as martyrs. For the victims of terror, on the other hand, they are brutal, ruthless and despicable perpetrators of violence, who must be responded to with counter-violence.

Terrorist behavior—from whatever direction it may come—represents an extreme form of ideology-driven aggression directed against people, institutions or buildings that are representative of a hated system. The ideological, political or religious-fundamentalist goals are to be achieved through intimidation, threat and the transmission of a terrorist message to the system being fought against [1]. If there are no motives of this kind, as for example in the case of the mass murder by means of firearms attack at a festival in Las Vegas in 2017, in which 58 people were killed and a further 851 injured [2], the act has to be classified as rampage or killing spree.

Beside rampages (personal motives, often mentally disturbed perpetrators), terrorism is to be distinguished from school shootings, actions by guerrillas, partisans or paramilitary organizations, and organized crime.

Terrorism has many faces:

- Ethnic and nationalist-separatist terrorism, which involves the struggle of ethnic or national minorities for autonomy;
- Anarchism, which is directed against representatives of disapproved forms of government;

This chapter is an expanded and updated version of an article that appeared in the journal *Der Nervenarzt* in 2020 [3]. Corresponding author: B. Bogerts.

- Left-wing terrorism, directed against a hated economic (capitalist), political and social system;
- Right-wing terrorism, guided by nationalist ideas and xenophobia;
- Religiously motivated terrorism. There are numerous examples of this in history for almost all major religions (see Chap. 20). Currently, the most significant religious form is Islamist/Salafist terrorism;
- State terrorism is practiced or supported by state organs in dictatorships. Extreme examples of this are Stalinism (fluctuating estimates of up to 60 million victims), National Socialism (up to 21 million victims) and Maoism (77 million victims) (see Chap. 18, Table 18.1).
- Terrorism by lone perpetrators who have usually invented their hate ideologies themselves or appropriated them from the Internet; like spree killers, they often suffer from psychiatric disorders.

Historical Background and Current Developments

Terrorist actions have remained a phenomenon of extreme politically or ideologically motivated acts of violence throughout history. Well-known historical examples are the Jewish resistance fighters against the Romans in the seventh century or the suicide attacks by Islamist assassins against political opponents or "infidels" in the eleventh century.

The course and manifestations of terrorism over the last 150 years have been divided into several successive periods [4].

The first phase is the anarchist phase, which lasted from the nineteenth century until the First World War. At the center of terrorism in this period were assassinations against representatives of state and Church, such as the murder of the Russian Tsar Alexander II in 1881, President Sadi Carnot of France (1894), King Umberto I of Italy (1900), President McKinley in the US (1901), or the assassination of Sarajevo, which eventually led to the First World War.

After the First World War, an anti-colonial wave of terror developed, beginning in the 1920s and continuing for about 40 years through independence and resistance movements, such as the Mau Mau struggles in Africa in the 1950s. The resistance movement of the Irish Republican Army (IRA) for the independence of Northern Ireland from Great Britain also falls into this period.

In the period from the 1960s to the 1980s, the third, predominantly left-wing terrorist wave emerged. The Red Army Faction (RAF) in Germany at this time pursued the goal of bringing down the capitalist system, which it declared to be the enemy, by terrorist attacks on representatives of the state and society. Other examples of this third wave of terror are the Red Brigades in Italy and the Nihon Sekigun (Red Army, RAF analogue in Japan). At the same time, in the 1970s and 1980s the Black Liberation Army, the Jewish Defense League, the Weather Underground and the Symbionese Liberation Army operated in the US.

This was followed by a fourth wave of terror, which has been sustained by Islamist jihadist motives up to the present day [4]. The content of Islamist terrorism includes the fight against the infidels (*kāfir/kuffār*) with reference to certain suras of the Koran as well as revenge for the disparagement of Muslim values by Western media. The first massive terrorist act of this jihadist wave was the attack on the World Trade Center on September 11, 2001; the terrorist attacks carried out or claimed by the Islamic State (IS) are also part of this, e.g., the series of attacks in Paris in November 2015, the attack on the Berlin Christmas market in December 2016 and the attack on a nightclub in Istanbul in January 2017.

Another wave of terrorism, right-wing terrorism, has grown in importance again in the last ten years, both by right-wing extremist terrorist groups and by individual perpetrators [5]. A prominent example of a right-wing extremist terrorist group was the National Socialist Underground (NSU) in Germany, which carried out ten racially motivated murders, 43 attempted murders, three explosive attacks and 15 robberies between 2000 and 2007. The most serious attacks by right-wing terrorists operating as lone perpetrators that have occurred so far took place in 2011 in Norway (Oslo and Utøya Island) with 77 deaths, New Zealand (Christchurch) with 51 and Texas (El Paso) with 22 deaths.

The Increasing Importance of the Internet

The structures of social media and the Internet offer an ideal breeding ground for a new form, the so-called "swarm terrorism," boosted by the anonymity of the Internet and the possibilities it provides of direct access to subcultural communities and the use of ideology-laden audio-visual media. Both self-radicalization through digital networking with ideologically like-minded people and strategically targeted radicalization by key terrorist figures take place [6].

After the most important social media companies in many countries were legally obliged to remove illegal content from their platforms in recent years, violent and violence-glorifying right-wing extremist online subcultures turned to alternative platforms to build digital hate communities. These are predominantly coined by right-wing extremist actors, whose numbers have increased in Europe in the last decade [7]. Popular means of communication include "memes," which are images or short films intended to convey right-wing extremist and racist content in an entertaining way. Memes often have an amusing, disparaging undertone mixed in to make it easier to attract the viewer to right-wing extremist and violent ideas [8]. The most common online topics of internationally networked right-wing extremist chat groups are hostility towards foreigners, migrants and Islam, conspiracy theories, white supremacist thinking, protective measures against supposed alienation and attacks on political dissenters.

It is mainly individual perpetrators who radicalize themselves through such media [9]. In this context, the Christchurch attack in March 2019 with 50 dead and just as many injured, represents a new category of international right-wing terror originating mainly on the Internet. The attacks in Poway (California, April 2019), El Paso (Texas, June 2019), Baerum (Norway, August 2019) and Halle (Germany, October 2019) can be classified along the same lines. All these perpetrators quote each other and invoke various versions of conspiracy or displacement theories against their own race, [10] just like the Norwegian mass murderer Anders Breivik (July 2011) before them.

Is Terrorism on the Rise?

After a steady rise in global terrorism with peaks in 2014 (33,555 terror deaths), there has been a decrease in terrorist attacks worldwide since the decline of the so-called Islamic State (IS) (Fig. 17.1). According to the Global Terrorism Index, 12,826 people died in terrorist attacks worldwide in 2019, [11] thus the number of global terror victims dropped by more than half between 2014 and 2019. However, the number of victims of terrorist attacks is still three times higher today than a decade earlier. In addition, far-right terrorism has increased in Western Europe, North America and Oceania for the fifth year in a row (250% increase in attacks between 2014 and 2019, 700% increase in fatalities) [11].

Right-wing extremist terrorist attacks with high numbers of victims have predominantly been committed by lone perpetrators, left-wing terrorist attacks predominantly by terrorist groups. The attacks in Vienna and France in 2020 show that Islamist terrorism is also still present in Europe. The Taliban in Afghanistan and Boko Haram in northern Nigeria, which is also active in the riparian states of Chad, Niger and Cameroon, were classified as the deadliest terrorist group worldwide.

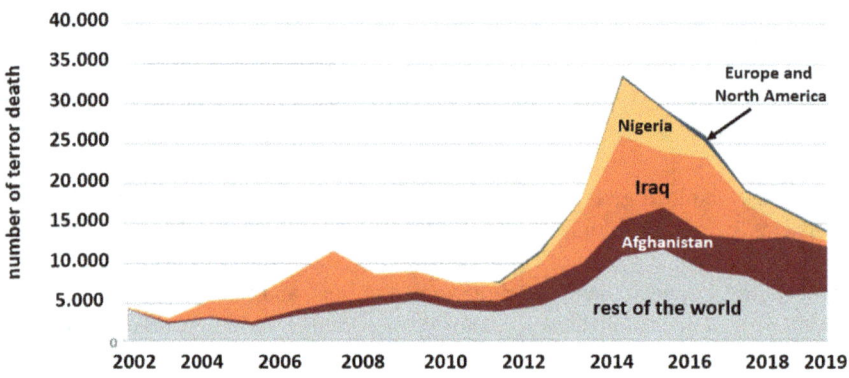

Fig. 17.1 Deaths from terrorist attacks worldwide, 2002–19. The increase from 2012 to 2015 resulted from the growing terrorist activities of the so-called Islamic State (IS). In the following four years, terrorist deaths declined by about 50%. The largest number of terrorist deaths occurred in Iraq, followed by Nigeria and Afghanistan. The numbers in Europe and North America were much lower. *Source* Institute for Economics & Peace. Global Terrorism Index 2020, Measuring the Impact of Terrorism [11]

The number of injured and psychologically traumatized victims is many times higher than the number of dead. The total economic damage in 2019 is estimated at US$26.4 billion. Afghanistan, Iraq, Nigeria and Syria were the most affected countries [11].

Who Becomes a Terrorist?

The political, social and psychosocial constellations are relevant for most inhabitants of a terror-affected region, but only an extremely small proportion of them become terrorists. What kind of people are they?

It is often argued that terrorist acts can only be carried out by people who are not psychologically normal, and that some mental disorder must be present that results in a willingness to radicalize and commit such acts of violence. Nevertheless, mental illnesses are only a secondary factor in the causal structure of terrorism [12, 13]. Reports on psychiatric illnesses in individual assassins or spree killers quickly give the impression that mentally ill people are generally more dangerous than mentally healthy people. It must be emphasized that the vast majority of people with mental disorders do not commit criminal acts. 95% of mentally ill people live violence-free lives, compared to 98% of people without mental disorders [13].

Terrorism is the most serious outgrowth of radicalization and extremism. The complex set of conditions resembles the multiple interacting causes of the general model of aggression shown in Fig. 12.5 (see Chap. 12). Pre-existing personality predisposition as well as violence-forming and traumatizing experiences in the early social environment during childhood and adolescence form characters with identity and self-esteem problems and an increased propensity to violence, which are susceptible to political or religious ideologies [14]. If political or social crisis situations and feelings of exclusion, disparagement or threat are added, then the way is paved for terrorist acts.

Psychology and Sociology of Group Terrorists

Left-Wing Terrorism

More detailed studies on the psychological, social and psychopathological conditioning factors of terrorist acts were first conducted in Germany about four decades ago on the occasion of the then dominant left-wing extremist terrorism of the Red Army Faction (RAF) [15–18]. The authors reported that left-wing extremist individuals, most of whom belonged to the RAF and the June 2 Movement, had an above-average social level of origin and for the most part had a university education with or without a degree [16]. One-third were women. In contrast, right-wing terrorism at that time, similar to today, was practiced exclusively by all-male groups with a pronounced cult of uniforms, weapons and fighting, and a lower average age (22 vs. 30 years). One-third of all terrorists had already committed juvenile offenses before joining the terrorist scene [16].

The largest proportion of students with later left-wing terrorist careers moved in ideologically homogeneous circles with strong pressure to comply, and lived in collective living arrangements such as communes. One-third of the left-wing extremists had a serious irreparable break with the parental home. Group membership became a psychologically existential question. Not the objective but the act of terrorist destruction itself provided the real satisfaction [15]. Psychiatric disorders were not found more frequently in terrorists than in the average population. Often, however, extremely pronounced personality traits were present, such as "neurotic hostility" [15]. It was argued that what terrorists have in common is an unfulfilled claim to the significance and power of young people who want to force society by means of violence to take note of their person and their will [15]. The authors found no indications for a psychopathological interpretation of terrorist actions [17]. Also, in several forensic psychiatric evaluations of accused left-wing terrorists no indications of court-relevant mental disorders could be found [18].

Right-Wing Terrorism

Conspiracy theories and xenophobia combined with ethno-national arrogance form the ideological core of all right-wing terror. The perpetrators of the attacks in Christchurch, Poway, El Paso, Baerum, Halle, as well as Utøya and Munich, invoked different variations of the world conspiracy theory of "great replacement" or "white genocide," which show extensive overlaps in content. The core of these theories, out of touch with reality as they are, is that the white European population is to be deliberately pushed back (often with accusations of supposed Jewish masterminds) and replaced by competing ethnic/religious groups (e.g., Muslims) whose culture is incompatible with the Western one. This narrative combines a wide variety of right-wing extremist, anti-feminist, anti-Semitic, anti-establishment and anti-migration ideologies [10]. In Germany, where it is also called *Großer Austausch* or *Volkstod*, ("great exchange" or "people's death"), similar ideas were already present in Hitler's *Mein Kampf* [19].

Paranoia (persistent psychiatrically relevant delusional symptomatology) and conspiracy thinking show a certain relationship, but are not identical phenomena. In paranoid-psychotic symptoms, the supposed threat or influence is always felt by the affected person to be directed only at himself. In conspiracy theories, a hostile attitude towards other groups of people is present from the outset. The perceived threat, even if out of touch with reality, is related by conspiracy theorists to the whole society or own group and not only to themselves. The reality-disturbed classification of supposed enemies against one's own group is the core idea of conspiracy belief and serves as justification for acts of terror [20].

The attraction of a conspiracy theory is all the greater, the more explanatory power and thus supposed control over accused grievances is gained. The attractiveness of world conspiracy theories that identify all-encompassing explanations and guilty groups thus becomes understandable. Conspiracy theories provide narratives for the devaluation of the foreign group and the upvaluation of the self-group, which do not require objective truths and are therefore particularly appealing to marginalized groups [21].

Right-Wing Terrorism as a Predominantly Male Phenomenon

With few exceptions, right-wing terrorism is a predominantly male phenomenon. Right-wing terrorist groups form a collective identity that is characterized by a so-called ingroup–outgroup (own-group–foreign-group) enmity. Group processes and a charismatic leader to whom the members are obedient are seen as important factors here. The traditional ideal of masculinity in the form of dominance, success, power, pain tolerance, endurance, independence, toughness and courage, which is embodied in the archaic image of the fighter, hero and victor, is claimed by many right-wing extremists for themselves [22, 23]. Strategies geared to this are used by terrorist groups and function via identification processes and feelings of group membership.

Islamist Terrorism

After the occupation of Iraq in 2003 by the US and the UK—with their trumped-up reasons for war—in pursuit of their geostrategic and economic interests, the Americans attacked Fallujah, the famous city of mosques, in 2004. The city was completely destroyed, and tens of thousands of civilians were killed or injured. The uranium munitions used by the US Army resulted in a dramatic increase in later secondary illnesses. During the same period, the torture scandals of Iraqi prisoners by US guards at the infamous Abu Ghraib prison occurred.

This situation and the ensuing chaos in Iraq encouraged the influx of terrorist organizations. The most significant of these became the so-called Islamic State (IS), which in its own turn made a name for itself through the most brutal excesses of violence. Attacks and suicide bombings were carried out by a large number of jihadist terrorists worldwide motivated by revenge for perceived humiliation and oppression, and in the defense of Islam.

A similar finding to that made for the left-wing terrorist scene of the RAF era—that there was no increased incidence of pathological psychosyndromes among group terrorists—has been made for more recent studies of Islamist terror. Several investigators in the United States and Europe found that the majority of Islamist terrorists had a college education or other academic or technical degree or vocational training

[24] and the same was true for suicide bombers [25, 26]. Perceived or actual discrimination and the perception that Islam was threatened by the Western world and therefore needed to be defended were the most common motivations for terrorist acts. The feeling of a threat from other cultures is thus a phenomenon shared by Islamist and right-wing terrorist groups. However, the vast majority of Islamist terrorists had neither a religious upbringing nor a profound knowledge of the Koran. In a summary of forensic psychiatric reports on 29 Islamist-motivated offenders, no psychopathological symptoms were found in 19 offenders who were immigrants to Germany, but dissocial conspicuousness was described; three had schizophrenic psychosis and two had a primarily dissocial problem [27].

Several international studies conclude that group terrorists are no more likely to suffer from a mental disorder of pathological value than people from comparable social backgrounds, thus there is no evidence that group terrorist behavior requires the presence of a psychiatric disorder [1, 28].

Other terrorism researchers report that it is not so much a disease-valent psychopathology as certain personality profiles that are associated with radicalization of terrorist groups. Narcissistic characters (self-conceit with extreme sickliness) and antisocial personality structures are mentioned, as well as a generally increased potential for aggression [13, 29, 30].

This is also supported by criminological analyses of groups of people who joined terrorist organizations. In 2016, the German Federal Office for the Protection of the Constitution presented an analysis of the radicalization backgrounds and trajectories of individuals who had left Germany for Syria or Iraq for Islamist reasons [31]. Two-thirds of those who left the country already had police records prior to their departure, particularly for violent or property-related offenses, as well as for politically motivated crimes and drug trafficking or possession. In the course of radicalization, there was an increase in politically motivated crime, followed by violent and property crime. More than half of these individuals had committed three or more criminal offenses, meaning that they were overwhelmingly multiple offenders. Pre-existing dissocial personality traits were common. Factors contributing to radicalization were mainly acquaintances from the Islamist scene, contacts in relevant mosques, the Internet, Islam seminars and Koran distribution campaigns. The same analysis further shows that approximately 80% of those who left the country were male, 20% female. The largest age group in terms of numbers was 22–25-year-olds. Among women, the desire for a life in a different/new Islamic social order and the marriage motive stood out. In contrast, the Islamist jihadist motivation was significantly more pronounced among men than among women [31].

It appears that when the whole spectrum of persons who left the country is considered, for persons who had been radicalized in the short term, the radicalization process took place less as a result of external stimuli, but apparently more as a "self-referential" process—a process that took place within the person without external influences. Short-term radicalized persons apparently had fewer points of contact with Islamist ideological content than long-term radicalized persons, for whom there were more frequent indications of possession of relevant Islamist propaganda material. Finally, it was concluded that latent willingness to use violence in connection with Salafist ideology leads relatively quickly not only to a fundamental affirmation of jihadist violence, but also to the desire to carry it out themselves. Returnees were therefore assumed to pose a particular security risk [31].

Comparable data on individuals who joined IS are also available from France, England, Italy, Belgium, the Netherlands and Greece [32, 33]. Between 50 and 70% already had a criminal record before joining the terrorist scene. One explanation for the high proportion of people with criminal records is that their tendency to aggressive and violent behavior is easier to act out in a terrorist environment, where it is less risky.

Special Features of Salafist Terrorism

Salafism per se is not to be equated with terrorism. However, Salafists claim true Islam, which refers to the founding generation, only for themselves and regard members of other religions or atheists as "infidels." Among violence-oriented groups, this can result in calls to kill, especially since extremist Salafists see Islam as threatened. Therefore, jihad, or holy war, becomes necessary. Jihadists accept their own deaths in order—as they believe—to defend Islam. In return, they are promised direct entry into paradise.

Surviving assassins reported that the longing for a heroic death and immortality is a strong motivation. Others spoke of the ecstasy of battle, which approached religious enthusiasm [24]. A document with prayers and reflections, intended to help the attackers persevere, was found on one of the perpetrators of the terrorist attack on the World Trade Center on September 11, 2001.

Undoubtedly, the personality traits outlined above also apply to many Salafist-inspired terrorists. On the other hand, it is probable that, in the case of sufficiently long and intensive indoctrination with radical Islamist or other ideologically influenced ideas since childhood and adolescence or after massive psychological traumatization through war, expulsion, destruction or oppression (see the example of Fallujah), a predisposed person can become a terrorist without the presence of an a priori existing psychological disturbance.

Common Characteristics of Terrorist Groups

The paths leading to violent radicalization have been described as quite diverse. Personal and collective crises following perceived or actual humiliation and exclusion, combined with a new sense of significance, belonging and obligation in the terrorist group, were the most important motives, albeit with varying degrees of intensity among individuals. In a study on this subject, the set of conditions leading to violent radicalization proved to be highly complex, with no uniform prototype [34].

The most important difference between radicalized individuals who did not become violent and those who did was what was seen in the latter as an emotional disposition to violent behavior, stimulation and "getting a kick" out of acts of violence, a terrorist code of honor with a pronounced "us versus them" narrative, and group pressure [29].

Rather than the presence of a manifest psychopathology of a delusional, schizophrenic, depressive or neurotic nature, members of terrorist groups appear to have a specific personality accentuation which can amount to personality disorder. Narcissistic, antisocial and paranoid-aggressive personality traits, emotional coldness, egocentrism, a feeling of emptiness and cold rationality are described [26, 35]. Nevertheless, terrorists have a broad spectrum of personality profiles with heterogeneous psychological characteristics as well as a complex structure of different psychosocial experiences, making it difficult to classify patterns or attribute particular characteristics to them [30].

From a social science perspective, the following types of perpetrators of extremist group violence have been distinguished [36]:

(1) Perpetrators who want to prove that they are opinion leaders. This type, which is extremely dominant in groups, sees acts of violence as an effective option for gaining recognition in the extremist context. Such perpetrators can be assigned to the histrionic (extremely needy for recognition) personality spectrum.

(2) Perpetrators whose central motive is affiliation with the group or its ideology. This dependent type looks for people in social contexts that provide security of orientation and behavior, and is therefore easily influenced. As a result of lived experiences with radical allies and indoctrination, a sense of increasing obligation to one's radicalized group develops. Such perpetrators most closely correspond to the type of a dependent personality.

(3) In the third type, an individual radicalization process takes place with a turn toward an ideology that makes violence appear justified, without initial contacts with a terrorist group organization, but subsequent integration into a violent collective.

For the development of Islamist terrorist attitudes, three essential psychosocial preconditions have been formulated [37]. First, an individual motivation that can result from humiliation, discrimination or oppression; second, an ideological narrative that is used to justify the use of violence; and third, a social network that spurs on the performance of radical acts.

Characteristic of all terrorist groups is a common image of the enemy and the need for cohesion in the face of high social pressure within the group. Dropouts run the risk of draconian sanctions including executions.

Characteristics of Lone-Wolf Terrorists

In contrast to terrorists who act in groups, mental illnesses are more prevalent in individuals who commit terrorist attacks as "lone wolves" [38, 39]. These perpetrators may well follow a specific ideology, which they have acquired primarily via the Internet, or may have been inspired by an existing terrorist organization. However, they are not subject to the hierarchical structures of a terrorist organization [40, 41]. These lone wolves are not always completely isolated, but are often in contact with like-minded people through hate forums on the Internet. Particularly with regard to lone wolves, a clear dominance of the male sex can be seen [42]. Furthermore, lone wolves are on average about ten years older than members of terrorist organizations [43].

The ideological background of lone-wolf terrorist attacks is not linked to a specific ideology. The entire spectrum of radical right-wing, radical left-wing and Salafist-jihadist ideas can be found among such perpetrators. Often, these terrorists also develop hate ideologies that stem from their own thinking and do not belong to any other extreme political or religious ideology.

Terrorist lone wolves are more likely to suffer from depression or paranoid symptoms than group terrorists [44]. They are more likely to live separately or in isolation with little or no social contact [38, 39, 43, 45]. Lone perpetrators were found to be 13 to 14 times more likely to have a mental illness than group perpetrators [39, 46]. These are primarily severe mental illnesses, particularly schizophrenic and delusional disorders as well as autism spectrum disorders [46]. Post-traumatic stress disorder, anxiety, depression and alcohol dependence have also been found to be clustered among lone offenders [47].

Severe mental illness is not typically associated with terrorism, but it appears to influence vulnerability to radicalization among lone wolves or to be a risk factor [13, 48]. Such terrorists are obviously not socially disadvantaged individuals. On average, their level of education corresponds to that of the general population, and they do not generally belong to lower income classes [49]. On the other hand, they are much more likely to be loners with only limited social contacts.

The presence of a mental disorder does not mean that such offenders are less able to carry out acts in a planned, orderly and effective manner. While these perpetrators often have social communication deficits and withdraw in the years before carrying out the crime, such social problems are not an obstacle to carrying out terrorist attacks with thorough planning and high numbers of victims. Often, in the run-up to such acts, there are life upheavals or experiences that those affected find dishonoring and belittling. The content of the ideology they have appropriated from others or developed themselves plays less of a role than its radicality.

When such acts are carried out, they are aimed at and prepared accordingly to achieve the highest possible number of victims; thus, for example, explosive devices, automatic weapons or even trucks are used to speed into a crowd of people.

Most Momentous Lone-Wolf Attacks in the Last Decade

July 2011, Oslo and Utøya Island, Norway
The 42-year-old right-wing terrorist Anders Breivik first killed eight people with a bomb in Oslo's government district before shooting 69 people on Utøya Island near Oslo. The mostly young victims were attending a summer camp of the youth organization of the Norwegian Social Democratic Labour Party. A psychiatrist initially diagnosed Breivik with paranoid schizophrenia. A second expert opinion did not consider Breivik psychotic, but diagnosed him with antisocial and narcissistic personality disorder [50].

June 2016, Orlando, Florida, USA
The 29-year-old perpetrator shot 49 people and wounded 53 others in a nightclub frequented mainly by homosexuals. He saw himself as an Islamic holy warrior. According to him, the act was a reaction to the death of IS leader Abu Wahib a few weeks earlier. He became violent towards his ex-wife and was described as homophobic, racist and sexist [51].

July 2016, Nice, France
The 31-year-old perpetrator drove a truck into a crowd on the seafront in Nice on a bank holiday. He killed 86 people and injured 458 others. The father of the perpetrator reported that he was depressed and consumed alcohol and other drugs. In addition, he had nervous breakdowns, during which he destroyed everything around him, screaming in anger. The crime was initially attributed to jihadist terrorism; however, concrete connections of the perpetrator to international Islamist terrorism could not be proven [52].

December 2016, Berlin, Germany
A Christmas market in the center of Berlin was the aim of this 24-year-old terrorist, who deliberately steered a truck into a group of people. Before doing so, he shot the driver of the truck. In total, he killed 12 people and injured 56 others. Before he came to Germany, he had served prison sentences in Tunisia and Italy. He continued his criminal career in the form of drug trafficking, assaults and identity fraud. At the same time, he radicalized himself in the Salafist-scene. Witnesses said that he went berserk with rage easily and blamed everyone but himself for his circumstances [53, 54].

January 2017, Istanbul, Turkey
In a nightclub where a New Year's Eve party was taking place, the 28-year-old perpetrator shot 39 people and injured 79 others. During the shooting he shouted *"Allahu Akbar"* (God is great). He was trained as a militia fighter in Afghanistan and Pakistan by Al-Qaeda and later by IS. The attack in Istanbul was apparently motivated by the fact that Turkey contributed to the downfall of the so-called Islamic State with military operations [55].

March 2019, Christchurch, New Zealand
The 28-year-old perpetrator shot 51 people in two mosques during Friday prayers and injured another 40. He posted his crime on the Internet via livestream on Facebook. Like the Norwegian mass murderer Breivik (July 2011), this right-wing terrorist believed in the conspiracy ideology of the "great replacement" and saw himself in a supposed war against the displacement of the white race. He acquired this racist ideology mainly in social networks and reinforced it through his travels and visits to European right-wing radicals [56].

How to Identify Lone Wolves?

Identifying lone wolves in advance of their acts is extremely difficult if not impossible, as their radicalization usually takes place in secret and is therefore not recognized by outsiders. The development of socially withdrawn, distrustful to hostile behavior, together with monitoring of Internet activities and messages posted on the Internet by such persons, as well as elaborate analysis of digital traces in social networks and the dark web, could provide clues to such risks. This could perhaps allow more intensive monitoring of possible perpetrators in some cases, or even enable them to be approached personally by psychiatrists or psychologists.

Returning Islamic State fighters may pose a particular threat, as they are trained to use firearms and explosives and are mentally still affiliated with terrorist networks. The death toll from returning terrorist fighters is many times higher than the death toll from perpetrators without combat experience [49].

The following criteria might indicate an increased risk of individual terrorist attacks: attempts or interest in traveling to regions where Islamist fighting takes place in order to be trained as a fighter; the length of time spent as a fighter in such conflict hotspots; other attempts to join a terrorist organization; and frequent visits to websites or chat forums with extremist content [49, 57].

Brain Structure and Brain Function of Terrorists

There have been no neuropathological studies of terrorists' brains to date, with the exception of the neuropathological study of the brain of Ulrike Meinhof, intellectual head of the Red Army Faction (RAF), which she joined in 1970 after undergoing brain surgery years earlier for a vascular tumor at the base of her brain. Until then, as a sought-after journalist, she used peaceful means to advocate the political goals she represented. After her suicide in 1976 in the high-security prison near the city of Stuttgart, a brain pathological examination was carried out. This revealed damage to the lower-middle temporal brain [58] caused by the earlier brain surgery (see Chap. 7, Fig. 7.3). This region is a central site of the limbic system responsible for the control of nerve cell groups whose activation results in aggressive behavior. It is therefore probable that the aggressive personality changes that became apparent in the years following her brain surgery were partly caused by the surgically induced brain damage. Of course, only the emotional extent of the aggressiveness can be explained by this, not the terrorist thoughts. These must be seen against the background of the overall political circumstances at the time.

Similar brain pathological findings have been reported in violent prisoners, including those who were convicted of planned, deliberately prepared (i.e., proactive) acts of violence, and who had personality traits that match characteristics as described in terrorists [58, 59].

Using structural and functional imaging techniques, brain studies of violent offenders who acted in a targeted and planned manner (proactive violence) found in some cases structural and functional deficits in the form of disrupted connections and reduced brain matter in the anterior and inferior frontal brain, limbic regions of the middle anterior temporal lobe (hippocampus, amygdala) and the associated cingulate cortex and insular region [58] (see Chap. 7, Fig. 7.1). These regions play a role in the neural modulation of emotions—including aggression—and largely correspond to brain areas activated during empathy and compassion [60].

However, no brain pathological or neuroradiological correlate could be found in the majority of the perpetrators of violence. This could be expected due to the multifactorial conditional framework of aggression and violence.

The male dominance in terrorist violence can also be attributed to brain-biological conditions. For example, the amygdala, which becomes active during aggression, is larger in men than in women. The volume of the inferior frontal cortex and the cingulate cortex, which are responsible for controlling the activity of the amygdala, is relatively smaller in males [61], which is why men are less able to keep negative emotions such as aggression in check (see Chap. 9).

Coincidence of Personality and Environment in Terrorists

The vulnerability-stress concept, according to which vulnerability (psychological susceptibility) is determined by the pre-existing (biographical, genetic, neurobiological) personality configuration, and the "stress" component by the current psychosocial, group-dynamic and political situation, also has a high explanatory value for the phenomenon of terror. According to this concept, even in the case of very low vulnerability in psychologically normal people, the strongest stressors can trigger acts of terror and, conversely, in the case of a pronounced personality disposition to acts of violence, minor occasions can suffice for this.

The political, religious, racist or other ideologies in whose name terror is perpetrated are hardly ever themselves the reason for terrorist acts of violence, but are actively sought out by many violence-prone individuals in order to act out their inclination and thereby achieve a feeling of superiority and significance or simply satisfaction through retribution or revenge for actual or perceived injustice.

A special pre-existing disposition is also required because, under the same psychosocial, political and ideological conditions, only an extremely small proportion of those affected become terrorists.

Group terrorism has always existed—albeit in different forms and different time cycles. The current world situation, especially that in the Near and Middle East and the right-wing extremism that is gaining ground again, does not suggest that anything will change in the foreseeable future. Nor is there any reason to assume that the number of lone perpetrators who carry out acts of terrorism, assassinations or rampages under paranoid notions or conspiracy theories will decrease.

The next attack is sure to come.

References

1. Post JM (2008) The mind of the terrorist: the psychology of terrorism from the IRA to Al-Qaeda, 1st edn. New York, NY: St. Martin's Griffin
2. Wikipedia.org (2017) Las Vegas shooting. Wikipedia.org. https://en.wikipedia.org/wiki/2017_Las_Vegas_shooting. Published 2021. Accessed April 1, 2021
3. Seidenbecher S, Steinmetz C, Möller-Leimkühler A-M, Bogerts B (2020) Terrorismus aus psychiatrischer Sicht, (Psychiatric aspects of terrorism). Nervenarzt 91(5):422–432. https://doi.org/10.1007/s00115-020-00894-0
4. Rapoport D (2002) The four waves of rebel terror and September 11. Anthropoetics 8(1):1–17. http://anthropoetics.ucla.edu/ap0801/terror/
5. Caniglia M, Winkler L, Métais S The rise of the right-wing violent extremism threat in Germany and its transnational character. ESISC—European Strategic Intelligence and Security Center. http://www.esisc.org/publications/analyses/the-rise-of-the-right-wing-violent-extremism-threat-in-germany-and-its-transnational-character. Published 2020. Accessed April 22, 2021
6. Baaken T, Schlegel L (2017) Fishermen or swarm dynamics? Should we understand Jihadist online-radicalization as a top-down or bottom-up process ? J Deradicalization 13:178–212. http://journals.sfu.ca/jd/index.php/jd/article/view/127
7. Guhl J, Ebner J, Rau J The online ecosystem of the German far-right. ISD—Institute for Strategic Dialogue. https://www.isdglobal.org/wp-content/uploads/2020/02/ISD-The-Online-Ecosystem-of-the-German-Far-Right-English-Draft-11.pdf. Published 2020. Accessed April 22, 2021
8. Bogerts L, Fielitz M. The visual culture of Far-Right Terrorism. PLOG Series Pandora. https://blog.prif.org/2020/03/31/the-visual-culture-of-far-right-terrorism/. Published 2020. Accessed October 8, 2020.
9. Jense M, James P, Lafree G, Safer-Lichtenstein A, Yates E The use of social media by United States extremists. National Consortium for the study of Terrorism and Responses to Terrorism. https://www.start.umd.edu/pubs/START_PIRUS_UseOfSocialMediaByUSExtremists_ResearchBrief_July2018.pdf. Published 2016. Accessed January 10, 2020
10. Davey J, Ebner J The great replacement, the violent consequences of mainstreamed extremism. Institute for strategic dialogue. https://www.isdglobal.org/isd-publications/the-great-replacement-the-violent-consequences-of-mainstreamed-extremism/. Published 2019. Accessed January 10, 2020
11. Institute for Economics & Peace. Global terrorism index 2020, Measuring the impact of terrorism. https://www.economicsandpeace.org/reports/. Published 2020. Accessed November 25, 2020

12. Bhui K, Everitt B, Jones E (2014) Might depression, psychosocial adversity, and limited social assets explain vulnerability to and resistance against violent radicalisation? Correa-Velez I, ed. PLoS One 9(9):e105918. https://doi.org/10.1371/journal.pone.0105918
13. Maier W, Hauth I, Berger M, Saß H (2016) Zwischenmenschliche Gewalt im Kontext affektiver und psychotischer Störungen (Interpersonal violence in the context of affective and psychotic disorders). Nervenarz. 87(1):53–68. https://pubmed.ncbi.nlm.nih.gov/26676656/
14. Beelmann A (2020) A social-developmental model of radicalization: a systematic integration of existing theories and empirical research. Int J Conf Violence. 14(1):1–14. https://doi.org/10.4119/ijcv-3778
15. Süllwold L (1981) Stationen in der Entwicklung von Terroristen: Psychologische Aspekte biografischer Daten. In: Jäger H, Schmidtchen G, Süllwold L (eds.) Lebenslaufanalysen. Wiesbaden: VS Verlag für Sozialwissenschaften; 9:79–116. https://doi.org/10.1007/978-3-663-14369-7_2
16. Schmidtchen G (1981) Terroristische Karrieren: Soziologische Analyse anhand von Fahndungsunterlagen und Prozeßakten. In: Jäger H, Schmidtchen G, Süllwold L (eds.) Lebenslaufanalysen. Wiesbaden: VS Verlag für Sozialwissenschaften 13. https://doi.org/10.1007/978-3-663-14369-7_1
17. Jäger H, Böllinger L (1981) Thesen zur weiteren Diskussion des Terrorismus. In: Jäger H, Schmidtchen G, Süllwold L (eds) Lebenslaufanalysen. VS Verlag für Sozialwissenschaften, Wiesbaden, pp 231–236
18. Rasch W (1979) Psychological dimensions of political terrorism in the Federal Republic of Germany. Int J Law Psychiatry 2(1):79–85. https://doi.org/10.1016/0160-2527(79)90031-1
19. Stanton GH QAnon is a Nazi Cult, Rebranded. genocidewatch.com. https://www.genocidewatch.com/single-post/2020/09/09/QAnon-is-a-Nazi-Cult-Rebranded. Published 2020. Accessed April 26, 2021
20. Van PJ, Douglas KM, De IC (2018) Connecting the dots: illusory pattern perception predicts belief in conspiracies and the supernatural. Eur J Soc Psychol 48(3):320–335. https://doi.org/10.1002/ejsp.2331
21. Douglas KM, Douglas KM, Sutton RM (2017) The psychology of conspiracy theories. Curr Dir Psychol Sci 26(6):538–542. https://doi.org/10.1177/0963721417718261
22. Möller-Leimkühler AM (2017) Why is terrorism a man's business? CNS Spectr 1–10. https://doi.org/10.1017/S1092852917000438
23. Möller-Leimkühler AM, Bogerts B (2013) Neurobiological, psychosocial and sociological condition. Nervenarzt 84(11):1345–1354, 1356–1358. https://doi.org/10.1007/s00115-013-3856-y
24. Sageman M (2004) Understanding terror networks. University of Pennsylvania Press
25. Post JM, Ali F, Henderson SW, Shanfield S, Victoroff J, Weine S (2009) The psychology of suicide terrorism. Psychiatry Interpers Biol Process 72(1):13–31. https://doi.org/10.1521/psyc.2009.72.1.13
26. Marazziti D, Veltri A, Piccinni A (2018) The mind of suicide terrorists. CNS Spectr 23(02):145–150. https://doi.org/10.1017/S1092852917000566
27. Leygraf N (2014) Zur Phänomenologie islamistisch-terroristischer Straftäter, (On the phenomenology of Islamic terrorist offenders). Forensische Psychiatr Psychol Kriminologie 8(4):237–245. https://www.researchgate.net/publication/288974608_On_the_phenomenology_of_Islamic_terrorist_offenders
28. Horgan J (2014) The psychology of terrorism, 2nd edn. Routledge, New York, NY
29. Bartlett J, Miller C (2012) The edge of violence: towards telling the difference between violent and non-violent radicalization. Terror Polit Violence 24(1):1–21. https://doi.org/10.1080/09546553.2011.594923
30. Piccinni A, Marazziti D, Veltri A (2018) Psychopathology of terrorists. CNS Spectr 23(02):141–144. https://doi.org/10.1017/S1092852917000645
31. Bundeskriminalamt (BKA), Bundesamt für Verfassungsschutz(BfV) HIK gegen E (HKE) Analysis of the background and process of radicalization among persons who left Germany to travel to Syria or Iraq based on Islamist motivations. bka.de. https://www.bka.de/SharedDocs/

Downloads/EN/Publications/Other/AnalysisOfTheBackgroundAndProcessOfRadicalization. html. Published 2016. Accessed April 26, 2021
32. Institute for Economics & Peace. Global terrorism index (2019) Measuring the impact of terrorism. Visionofhumanity.org. https://www.visionofhumanity.org/wp-content/uploads/2020/11/GTI-2019-web.pdf. Published 2019. Accessed January 10, 2020
33. Campelo N, Oppetit A, Neau F, Cohen D, Bronsard G (2018) Who are the European youths willing to engage in radicalisation? A multidisciplinary review of their psychological and social profiles. Eur Psychiatry 52:1–14. https://doi.org/10.1016/j.eurpsy.2018.03.001
34. Jensen MA, Atwell Seate A, James PA (2018) Radicalization to violence: a pathway approach to studying extremism. Terror Polit Violence 1–24. https://doi.org/10.1080/09546553.2018.1442330
35. Marazziti D (2016) Psychiatry and terrorism: exploring the unacceptable. CNS Spectr 21(02):128–130. https://doi.org/10.1017/S1092852916000031
36. Pisoiu D, Zick A, Srowig F, Roth V, Seewald K (2020) Factors of individual radicalization into extremism, violence and terror—the German contribution in a context. Int J Conf Violence 14(2):1–12. https://doi.org/10.4119/ijcv-3803
37. Webber D, Kruglanski AW (2018) The social psychological makings of a terrorist. Curr Opin Psychol 19:131–134. https://doi.org/10.1016/j.copsyc.2017.03.024
38. Gruenewald J, Chermak S, Freilich JD (2013) Distinguishing "loner" attacks from other domestic extremist violence. Criminol Public Policy 12(1):65–91. https://doi.org/10.1111/1745-9133.12008
39. Corner E, Gill P (2015) A false dichotomy? Mental illness and lone-actor terrorism. Law Hum Behav 39(1):23–34. https://doi.org/10.1037/lhb0000102
40. Alakoc BP (2015) Competing to kill: terrorist organizations versus lone wolf terrorists. Terror Polit Violence 0:1–24.https://doi.org/10.1080/09546553.2015.1050489
41. Bakker E, de Graaf B (2010) Lone wolves: how to prevent this phenomenon? Terror Counter-Terrorism Stud 1–8.https://doi.org/10.19165/2010.2.01
42. Lauritsen JL, Heimer K, Lynch JP (2009) Trends in the gender gap in violent offending: new evidence from the national crime victimization survey. Criminology 47(2):361–399. https://doi.org/10.1111/j.1745-9125.2009.00149.x
43. Gill P, Horgan J, Deckert P (2014) Bombing alone: tracing the motivations and antecedent behaviors of lone-actor terrorists. J Forensic Sci 59(2):425–435. https://doi.org/10.1111/1556-4029.12312
44. Hewitt C (2003) Understanding terrorism in America. Routledge, New York, NY
45. Spaaij R (2012) Understanding lone wolf terrorism: global patterns, motivations, and prevention. Springer, New York
46. Corner E, Gill P, Mason O (2016) Mental health disorders and the terrorist: a research note probing selection effects and disorder prevalence. Stud Confl Terror 39(6):560–568. https://doi.org/10.1080/1057610X.2015.1120099
47. Alderdice TL (2007) The individual, the group and the psychology of terrorism. Int Rev Psychiatry 19(3):201–209. https://doi.org/10.1080/09540260701346825
48. Borum R (2014) Psychological vulnerabilities and propensities for involvement in violent extremism. Behav Sci Law 32(3):286–305. https://doi.org/10.1002/bsl.2110
49. Pantucci R, Ellis C, Chaplais L (2015) Lone-actor terrorism. Lit Rev Royal United Services Institute for Defence and Security Studies, London
50. Breivik AB Wikipedia.org. https://en.wikipedia.org/wiki/Anders_Behring_Breivik. Published 2021. Accessed April 1, 2021
51. Mateen O Wikipedia.org. https://en.wikipedia.org/wiki/Omar_Mateen. Published 2021. Accessed April 1, 2021
52. Lahouaiej-Bouhlel M Wikipedia.org. https://en.wikipedia.org/wiki/Mohamed_Lahouaiej-Bouhlel. Published 2021. Accessed April 1, 2021
53. Wikipedia.org (2016) Berlin truck attack. Wikipedia.org. https://en.wikipedia.org/wiki/2016_Berlin_truck_attack. Published 2021. Accessed April 22, 2021

54. Amjahid M, Müller D, Musharbash Y, Stark H, Zimmermann F An attack is expected. zeit.de. https://www.zeit.de/politik/deutschland/2017-04/berlin-attack-christmas-market-breitscheidplatz-anis-amri/komplettansicht. Published 2017. Accessed April 22, 2021
55. Wikipedia.org. Istanbul nightclub shooting. Wikipedia.org. https://en.wikipedia.org/wiki/Istanbul_nightclub_shooting. Published 2021. Accessed April 1, 2021
56. Wikipedia.org. Christchurch mosque shootings. Wikipedia.org. https://en.wikipedia.org/wiki/Christchurch_mosque_shootings. Published 2021. Accessed April 1, 2021
57. Lindekilde L, O'Connor F, Schuurman B (2019) Radicalization patterns and modes of attack planning and preparation among lone-actor terrorists: an exploratory analysis. Behav Sci Terror Polit Aggress. 11(2):113–133. https://doi.org/10.1080/19434472.2017.1407814
58. Bogerts B, Schöne M, Breitschuh S (2018) Brain alterations potentially associated with aggression and terrorism. CNS Spectr 23(2):129–140. https://doi.org/10.1017/S1092852917000463
59. Schiltz K, Witzel JG, Bausch-Hölterhoff J, Bogerts B (2013) High prevalence of brain pathology in violent prisoners: a qualitative CT and MRI scan study. Eur Arch Psychiatry Clin Neurosci 263(7):607–616. https://doi.org/10.1007/s00406-013-0403-6
60. Singer T, Klimecki OM (2014) Empathy and compassion. Curr Biol 24(18):R875–R878. https://doi.org/10.1016/j.cub.2014.06.054
61. Bogerts B, Möller-Leimkühler AM (2013) Neurobiological and psychosocial causes of individual male violence. Nervenarzt 84(11):1329–1344. https://doi.org/10.1007/s00115-012-3610-x

Chapter 18
Collective Violence, Xenophobia, Pogroms, Genocide

B. Bogerts and C. Steinmetz

The greatest catastrophes of mankind have been caused by collective violence in the form of wars and genocide. In addition to a brief description of their historical dimensions, this chapter offers possible explanations for the occurrence of such events from phylogenetic, social science, mass psychology and neuroscience perspectives.

Collective Violence as a Legacy of Evolution

Collective violence by groups, clans, religious or ethnic communities, bands, tribes, peoples or states against others not belonging to the own group is as old as mankind itself. It is not the result of only recent developments of social or political conflicts, although these can often play a decisive role as triggering mechanisms. Rather, it is a disposition for a spectrum of behaviors that has evolved over hundreds of thousands of years of human development and that has prevailed in phylogenesis because groups that were successful in collective violence had a survival advantage and were more likely to pass on their mentality than those that were inferior (see Chap. 4). The predisposition to collective violence, like that to individual violence, is present as an archaic vestige in all humanity as an inherited negative collective mortgage, even if most are so unaware. It is a phenomenon that can be assigned to the area of the collective unconscious and regularly reappears in certain group-dynamic or social or political constellations [1].

Similarities Between Humans and Animals

Group aggression against own conspecifics has not only developed in the course of evolution in *Homo sapiens*, but is also widespread in the animal kingdom. Group-living predators such as wolves, hyenas and lions engage in territorial fights against other packs; the same is true for some monkey species, mongooses, neighboring rat tribes and even ants [2, 3]. Such behaviors are particularly well described in our closest biological relatives, the chimpanzees. When one horde of male chimpanzees encounters another, attacks, often fatal, are launched by the outnumbered group, apparently after assessing the attack's chances of success [4–6].

Phylogenetic similarities in the evolution of group aggression in humans and apes are also indicated by surprising similarities in the behavior of juvenile gangs and chimpanzee groups. The groups of both species consist predominantly of male members and have a fixed hierarchy; status within the group is determined by the ability to fight as well as by skill in forming alliances; and within the group, violence escalates when hierarchy is unclear. Attacks on neighboring territories or groups occur regularly, especially if the rivals are considered weaker [7].

Humans and apes separated from their common precursors about seven million years ago. The similarities of primitive group-aggressive instincts, which still exist today, lead us to assume that these instincts had already shaped the interaction of prehumans and predecessors of *Homo sapiens* for hundreds of thousands of years and had a decisive influence on the course of human history even in prehistoric times. This is also an explanation for the fact that the precursors of today's man, such as *Homo erectus*, *Homo habilis* and Neanderthal man, disappeared in the course of human evolution; they were displaced by more intelligent, more agile, stronger and better organized human races that were thus more successful in collective violence.

The fact that warfare was part of the repertoire of prehistoric hunter-gatherers is proven by numerous excavations. On the shores of Lake Turkana in Kenya, numerous 10,000-year-old skeletons were found lying together, showing the impact of blunt impact weapons. At a site in Sudan, remains of 59 people from the period 14,000 to 12,000 BCE were found with flint points in their skeletons [8]. In Europe, too, there were a number of finds of massacres from the Neolithic period, proving that there was collective violence on a large scale here as early as 7000 years ago (see Fig. 18.1 and Chap. 4, Figs. 4.1–4.3).

Since the beginning of historical records kept several thousand years before our era by the ancient Egyptians, Babylonians, Assyrians, Persians, as well as in the Old Testament, in Greek and Roman historiography, warlike subjugation or extermination of other peoples and expansion of the power claims of the respective rulers are depicted as determining the course of world history. It was not different in prehistoric times. Collective violence was also present in Central and Southern Europe at that time, even if not reported by historians (Fig. 18.1). In the Tollense Valley in Mecklenburg-Vorpommern, Germany, a battlefield dating back to 1300 BCE has been excavated with up to more than 1000 dead. Genetic studies of the skeletal remains revealed that they were the remains of two different groups of people [10].

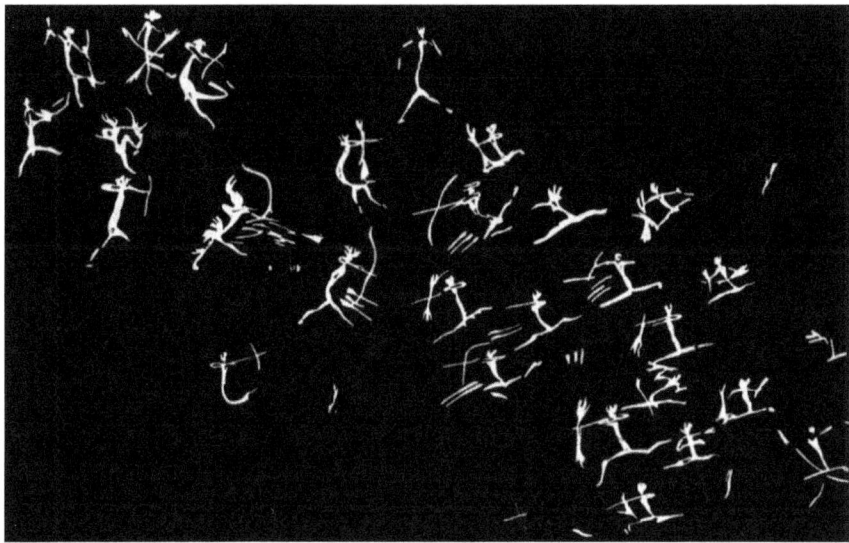

Fig. 18.1 Battle scene, Neolithic period, ca. 5000 BCE, les Dogues, Spain. *Source* Meller and Schefzik [9]; with permission of the Landesamt für Denkmalpflege und Archäologie Sachsen-Anhalt; see Ref. [9]

Historical Dimensions of Collective Violence

Our disposition to collective acts of violence, which can be deduced from our history through phylogenetic regularities and which can also be proven for prehistoric times, also unfolded to the full extent in historical times. The historiography is dominated less by reports of positive achievements and cooperative behavior of mankind than by battles, subjugations, uprisings, victories, expulsions and annihilation of the vanquished. The question has always been, and still is, who claims the future for himself. The rise and fall of all the great and small empires in world history bear witness to this.

The self-portrayals of ancient rulers carved in stone and ancient historiography, beginning with Homer (eighth century BCE) and Thucydides (460–395 BCE) as well as the later course of the history of Europe, Asia, Africa and America is characterized by wars and genocides. As a phylogenetic legacy, these have remained a constant worldwide phenomenon to this day, again and again disrupting the predominantly peaceful coexistence of humankind.

An impressive overview of the consequences of violence, terror, genocide and war only in the twentieth century was compiled by the American historian Rummel [11, 12]. According to this, more than 200 million people died worldwide in the last century as a result of state terror against its own population or wars against neighboring states (see Table 18.1). These included victims of the regimes in Cambodia (1975–1979), China (1949–1978), colonialism, Nazi Germany (1933–1945), Japan

Table 18.1 Consequences of state terror and genocides in the twentieth century. Only events with more than one million deaths are listed

Country	Period	Fatalities
China (Mao)	1949–1978	77,277,000
USSR	1917–1987	61,911,000
Colonialism	1900–Indep.	50,000,000
Germany	1933–1945	20,946,000
Congo (Belg.)	1885–1908	10,000,000
Japan	1936–1945	6,000,000
Cambodia	1975–1979	2,035,000
Turkey	1909–1918	1,883,000
Vietnam	1945–1987	1,670,000
Poland	1945–1948	1,585,000
North Korea	1948–1987	1,563,000
Bangladesh	1958–1987	1,500,000
Yugoslavia (Tito)	1944–1987	1,072,000

Source Data from R. J. Rummel, Death by Government (1994) [11] and Statistics of Democide (1997) [12]

during the Second World War, Pakistan (1958–1987), Poland in the years after the war (1945–1948), North Korea (1948–1987), Turkey (1909–1918), Vietnam (1945–1987), and Russia/USSR during both world wars and at the time of Stalinism, with nearly 140 million deaths cited by Rummel for Russia and China combined for this period alone.

Even if these figures can only be rough estimates and are questioned by some historians [13], they give an impression of the huge dimensions of state terror and genocide that persisted into modern times.

If one looks at the immense numbers of victims of wars and genocide in the twentieth century, one gets the impression that there has been a constant increase in individual, but especially collective, violence in world history. However, this view has been challenged by the American psychologist Steven Pinker in his influential 2011 book *The Better Angels of our Nature—Why violence has declined* [14]. Pinker concludes that violence has declined in the course of human history up to the present day and that today—at least in the Western world—we are living in the most peaceful period since the beginning of mankind. Pinker divides the course of world history leading to a constant reduction of violence into several successive phases [15]. The first phase was the transition from the society of hunter-gatherers, characterized by anarchy, about 10,000–5,000 years ago, to advanced agricultural cultures with cities and governments. The second phase took place in the late Middle Ages, associated with a decline in chivalric feuding and increasing state monopoly on the use of force. The third phase unfolded in the seventeenth and eighteenth centuries, thus in the age of the European Enlightenment, when torture and sadistic punishments were increasingly socially outlawed, combined with the rise of humanism. The final phase of violence reduction took place after the end of the Second World War. Since then, with the exception of the wars in the successor states of the former Yugoslavia and those in Eastern Ukraine, there have been no more wars in Europe or between Western industrial societies. Since the Universal Declaration of Human Rights in 1948, there has been—according to Pinker—an increasing rejection of violence against ethnic minorities or other political groups.

However, in regions of the world where the civilization process took place to a lesser extent or not at all, murder rates and group violence remain high. At present, murder rates in several Central American states are roughly equivalent to those of the European Middle Ages. Even among indigenous peoples in inhospitable regions of the world, everyday violence is many times more common than it is in Europe.

The lessons of the Nazi era and Stalinism were not learned in the same way in all regions of the world. State terror and genocides of immense proportions continued after the Second World War not only in several Asian countries (e.g., China, Vietnam, Cambodia, North Korea, the Philippines, East Pakistan), but also in Africa. Conflicts within and between African states and ethnic groups (Sudan, Rwanda, Burundi, Congo, Uganda, Angola, Ethiopia, Somalia) killed a total of more than 5 million people between 1955 and 2001 [16]. Since the end of the Second World War, 60 million people are believed to have fallen victim to genocides and "ethnic cleansing" operations in totalitarian-ruled communist regimes alone [17]. In addition, there were an estimated further 10 million deaths due to attempts by colonial powers to maintain control over occupied colonies, as well as the expulsion and elimination of Germans from Eastern European countries previously occupied by the Wehrmacht.

The conditions of past centuries were by no means better. According to Pinker [14], in the eighth century the An Lushan rebellion against the Tang dynasty in China, with 36 million dead, had more victims in relation to the world population at that time than all later warlike catastrophes. The Mongol conquests under Genghis Khan in the thirteenth century were similar: all cities and peoples in Asia and Europe that did not surrender to him were destroyed, with an estimated death toll of 40 million. Comparatively devastating were the consequences of the 30 Years' War in the seventeenth century; nor should we forget the several hundred thousand victims, including starvation deaths, of the Indian Rebellion against British colonial rule in the nineteenth century [18]. Wherever Europeans encountered militarily inferior civilizations in other continents in past centuries, these were subjected to total colonial despotism. The indigenous peoples of North, Central and South America were almost exterminated.

Risk Factors for Wars and Genocides

There are a number of approaches to identifying risk factors for genocides. It has been found that such events almost always occur in the wake of the collapse of a state, after a civil war, a revolution or a *putsch* [16], thus in conditions of anomie [19], during which assertions of power have not yet been consolidated, hitherto valid rules and laws no longer apply and new democratically legitimized state monopolies of violence are unable to act. Functioning democracies appear to be a decisive factor in preventing war and genocide. Democratic states are less likely to wage war against each other and less likely to be the site of civil wars and genocides [11]. There is also evidence that states that are more intensively interconnected by international trade have a lower incidence of war and genocide [12, 17].

Other risk factors for genocide that are particularly relevant to the present have been defined as [16, 20]:

- state-tolerated discrimination against ethnic or religious minorities,
- ideologies of a ruling class that exclude other ethnic groups,
- exercise of monopoly of power by an ethnic minority,
- dictatorships and autocracies,
- persecution of minorities and genocides in the past,
- political instability and civil wars,
- low level of integration in world trade.

It is often considered that ideologies such as National Socialism, fascism, communism and Islamism are the root causes of genocides. However, ideologies do not arise out of themselves, but are conceived in the minds of concrete people and adopted by individuals who use one or another ideology to act out claims to power, expansionist desires, delusions of grandeur or sadistic character traits without scruples. Examples from more recent times are dictatorial regimes and autocrats who were the cause of mass murders of unimaginable dimensions: Hitler, Stalin, Mao Zedong, Pol Pot, Idi Amin.

Sociological Studies on the Emergence of Group Hatred and Violence

Xenophobia, group hatred and the resulting collective aggression and violence always arise spontaneously anew due to our phylogenetically derived disposition for this, if they are not prevented by education, upbringing and cultural measures that inhibit emerging xenophobic attitudes and instead promote prosocial behaviors. A series of social-scientific and social-psychological experiments has demonstrated the regular spontaneous development of superiority thinking of one's own group, accompanied by devaluation of foreign groups and resulting collective aggression.

Social Identity

The spontaneous behavior of one group of people toward another was first studied from a social science perspective by Henry Tajfel (1919–1982) (Fig. 18.2). Tajfel, who himself had lost Jewish relatives through the Nazi terror, asked himself how the arrogant behavior of one group of people toward another can be explained and whether there is a spontaneous disposition to this in the nature of man. He recruited two groups of young men who differed only in that the members of one group liked paintings by the artist Paul Klee better, while those of the other group preferred paintings by the artist Wassily Kandinsky [21–24]. The advantage of this supposedly trivial group distinction was that the groups were not influenced by existing prejudices, previous conflicts, hierarchies or competition. Tajfel analyzed how the behavior of group members toward each other and their attitudes toward the other group evolved over time. He found that as soon as people feel they belong to a group, they perceive more positive characteristics in their own group and more negative characteristics in the foreign group [25]. Individual group members become carriers of collective characteristics in a positive and negative sense [23, 26]. It is important to be part of the "better" group because this has a positive effect on one's own self-esteem. Therefore, the foreign group quickly becomes the target of discrimination, since this supposedly raises one's own group, one's own social identity and thus ultimately one's own self-esteem. The actual characteristic on which the groups differ is not important. In his "theory of social identity," Tajfel showed that conflict can arise solely on the basis of belonging to a group.

The evaluation of foreign groups does not always have to be negative. Interest in and curiosity about what they do and perhaps do better, even the attitude that their way of life is more advantageous and acceptable and at least parts of it could be adopted, is as common as group arrogance [27]. A prerequisite for such an attitude, however, is that there is a certain level of education in one's own group.

Henry Tajfel
(1919 – 1982)

Fig. 18.2 Henry Tajfel (1919–1982). Portrait d'Henri Tajfel; European Association of Social Psychology, created on 25/02/2012; Available online at: https://de.wikipedia.org/wiki/Henri_Tajfel#/media/Datei:Henri_Tajfel.jpg; (retrieved October 2020)

Established and Outsiders

Another study on the mutual segregation of groups was conducted in an English suburb with a population of about 5000 [28]. Two groups were studied whose only distinguishing feature was that the established group had lived in the neighborhood for some time and the outsider group consisted of newcomers. Many even worked within the same factories. Despite this sameness, the established stigmatized the outsider group by labeling them as debauched and inferior. Likewise, members of the outsider group were discouraged from holding relevant positions, such as within the city council, churches or clubs [28]. An important component of this stigmatization was identified as a *pars pro toto* bias, i.e., a sweeping generalization of group characteristics that describes the same mechanisms as were highlighted by Tajfel. The worst characteristics of a member of the outsider group are portrayed as characteristic of that group, and the best characteristics of a few members of the establishment are seen as typical of the entire establishment group. This should ensure the dominance of the established.

Realistic Group Conflicts

On the question of the extent to which conflicts between groups trigger violence, in an early experiment conducted in Oklahoma, USA, boys of the same age were divided into two summer camp groups of equal size [29]. The groups were largely identical in terms of their family, socioeconomic and psychological backgrounds. In the first phase of the experiment, the adolescents got to know each other within their own group, made friends, made flags for their group, and thus gradually identified more and more with their own summer camp group. During the second phase of the experiment, the two groups, which had previously had no contact with each other, were pitted against each other in playful competitions to win prizes. After a short time, hostility developed between the groups, which began with insults and prejudices and culminated in the burning of the flag of the opposing team and the theft of private belongings, so the experiment had to be stopped. However, the enmity between the groups remained. Only within the third phase, in which the two groups had to achieve common goals together, did the hostility decrease. A comparable study was also abandoned due to escalating fights between the rival youths [29, 30]. These experiments show that conflicting goals, especially competition for resources between groups, involve a high risk of mutual aggression.

Felt Threat

Actual or perceived threat or disparagement of the values of one's own group can be a powerful trigger of collective violence [31]. A study on this in several countries in Europe, the US and Asia investigated whether perceived threat influences the propensity to violence of people belonging to a particular religious group [32]. Regardless of the country, it was found that as identification with the values of one's own group increased, the feeling of threat also increased and more violent intentions were expressed toward the foreign group. This effect was repeated in follow-up experiments. Videos were shown in which members of another social group spoke pejoratively about the group under study and portrayed their lifestyle as incompatible with their own (symbolic threat). Compared to groups of the same religion that did not watch these videos, the study group was significantly more likely to report supporting violence against and even participating in violent political persecution of the offending video group [33]. For the interpretation of these experiments, it is important to note that these group threats in the form of a video representation were merely virtual in nature. Whether they were real at all or factually verifiable did not play a role in the readiness to use violence that was elicited.

Group-Based Misanthropy: Right-Wing Extremism

Overvaluation of one's own group, devaluation and exclusion of strangers, asylum seekers, migrants and war refugees, Muslimophobia, anti-Semitism, racism, antiziganism, and devaluation of unemployed or disabled people are characteristic of right-wing extremist attitudes. Extensive research has been conducted in Germany on the prevalence of such group-based hostility [34, 35]. Devaluation of asylum-seeking and long-term unemployed people was present in more than half of the respondents in 2018–2019; xenophobia, Muslimophobia, Israel-related anti-Semitism, devaluation of Sinti and Roma in about 20% of the respondents. Xenophobia was about equally prevalent among both genders, and more pronounced among older citizens and those with lower levels of education [34].

In the representative survey study in 2019, 5.5–6.5% of the German population could be identified as openly approving of violence [34]. There was a close correlation between group-based misanthropy, right-wing extremist attitudes and advocacy of violence. For example, two-thirds of the respondents with a propensity for violence exhibited devaluation of asylum-seekers. The results of the study are remarkable because the respondents were interviewed in peacetime in a country with a functioning legal system, high prosperity and, due to its history, a sensitized attitude toward racism and the devaluation of social groups. Right-wing xenophobic crime represented by far the highest proportion of politically motivated crime in Germany in 2020 [36]. This situation is likely to be similar in other European countries.

Fig. 18.3 Evolutionary-biologically determined herd instinct to aggressive behavior combined with euphoric feelings of superiority. Partial image on left from DPA-Picture-Alliance No.: 31568260; partial image on right; Deposit photo No.: 152380326

Group Violence as a Male Domain

Collective as well as individual physical violence is largely practiced by male individuals. This is not only true for *Homo sapiens*, but seems to be a general biological principle, ultimately due to the Y chromosome and its effect on hormone balance, body structure and brain function. However, masculinity is not only a biological-anthropological fact, but has also grown as a historical-societal construction. While this may vary culturally, it has led to the establishment of a patriarchal order in most societies. Hegemonic (i.e., dominance-oriented) masculinity is focused on power and status, to which everything else must be subordinated. Despite today's changing gender roles, it continues to be seen as the action-guiding ideal for large parts of the male population, illustrated in particular in the rocker, hooligan and gang scenes as well as in right-wing extremist milieux—but not only there [1]. Violence functions in this context as a group norm, as a means of demonstrating masculinity and superiority, of securing status and raising self-esteem by attacking other, usually weaker groups. A comparison with the male chimpanzee hordes described above is obvious (Fig. 18.3).

In the case of hooligans and other "sense-making" male fighting communities, a preference for violence as a behavioral option is always also a symptom of deficient socialization and the failure of regular interpersonal communication opportunities. Violence thus becomes a kind of "universal substitute language" which, as an everyman resource [37], offers a catch-all lifeworld to those who do not have access to other ways of communicating in order to raise their self-esteem [38].

Suspension of Inhibition Mechanisms: Behavior in War

The conditions under which brutal violence by the average man can come about were impressively analyzed in the American historian C. R. Browning's book *Ordinary*

Men... [39]. Browning describes how in 1942 the Hamburg Police Battalion 101, consisting of 500 men, was sent on a "special assignment" to Poland. The men of this police battalion had all been well integrated as citizens up to that point, had families and in particular had not attracted attention by having committed any criminal offenses. Once they arrived in Poland, they were told that they had to carry out the extermination of Polish Jewry, with National Socialist ideologies and slogans being used to motivate them. The battalion commander informed the troops before issuing the order to deploy that those who did not feel up to the task could step forward to be assigned to another task in the battalion. Of the 500 police officers in the battalion, only 12 stepped forward. All the others participated in the rounding up and subsequent execution of the victims. The fact that these "completely ordinary men" were able to kill thousands was explained by the assumption that, on the one hand, an ideology glorifying and enabling violence, on the other hand, the compulsion to behave in conformity with the group, and, in addition, the handing over of personal responsibility to others (orders from above) overrode the ethical/moral norms and the associated psychological mechanisms inhibiting violence [40]. The overwhelming majority of the perpetrators returned to their completely normal, inconspicuous lives again after the end of the Nazi era.

Unique insights into soldiers' attitudes toward fighting, killing, and dying during the Second World War are provided by extensive analyses of Allied wiretap transcripts of German prisoners [41]. The authors conclude that people base their actions less on abstract ideologies than on what is expected of them by the group in which they find themselves. Since it is difficult to orient oneself in wartime situations, the comradeship group and its actions become the reference point for orientation. The military value system and the nearby social world of the soldier are decisive for his actions. However, the National Socialist extermination actions during the war and the Holocaust also showed that the majority of SS men, soldiers and police officers behaved violently when it was demanded or presented as appropriate. Within the overall context of events in the Third Reich and the accompanying intensive all-surpassing National Socialist propaganda, there was a shift in the "frame of reference" [41] from the civilian state to that of war, which justified the extermination of marginalized groups of people and those declared to be inferior. Racism and anticommunism were a political program against which even the vast majority of the intellectual elite at universities and faculties did not protest. The classification of the common man into the "master race" raised his self-esteem and made him conform to the behavior of the "master race." Many soldiers reported that whether as fighter pilots, in the navy or as commandos on land, they enjoyed shooting at people and killing for the sake of killing. Many soldiers did not need to become accustomed to the use of violence once they began their deployment, as this was part of their frame of reference in which killing became a duty [41]. This marital displaced "frame of reference" at least partially explains the crimes against civilians, including women and children, that are the rule rather than the exception in almost all armed conflicts worldwide.

Disinhibition as a Phenomenon of Mass Psychology

Another explanation for the perpetration of collective violence by otherwise normal people is to be found in the psychology of crowds. Everyone is familiar with the phenomenon of mood transference, which is found not only in animals living in packs but also in humans. The transmissibility of moods promotes group behavior in the same direction; in the course of evolution, this contributed more to the survival of a group than individual actions of many single persons because of the bundling of forces for the accomplishment of common tasks.

Mood transfer in the positive sense takes place, for example, at folk festivals, sporting and musical events, meetings in a party mood and at carnivals; negative variants are, for example, mass panic, yobbish behavior, riots and pogroms. That moods and emotions can be contagious is also true for collective aggression and violence. This not only applies to hedonistic group violence by hooligans and gangs, but also explains the celebratory mood found in many European countries at the beginning of the First World War.

In the anonymity of larger crowds, many people behave differently than they would as individuals. They allow themselves to be carried away by mood transference and set aside personal responsibility and inhibitions because that's what others also do. In a crowd, individuals can achieve a sense of significance and act out emotions that they would have to curb as individuals [42, 43]. The best example of this is the behavior of fans in soccer stadiums and of participants in mass protests. Collective violence can have a hedonistic component and is sought out by people who are disposed to do so. This is indicated by the high attendance figures at martial arts events, the behavior of hooligan and "ultra" groups who meet outside soccer stadiums on the "field" to beat each other up [44], the activities of riot tourists who travel from place to place to confront the police, and also by individual actions of troop units and paramilitaries in conflict regions around the world (see Chap. 13). Previously valid ethical and moral norms are invalidated when the reference group or community that created and transmitted such norms no longer considers them relevant and replaces them with new ideologies that condone violence. The ground is then prepared for the spillover of violence.

Brain Biological Correlates of Group Aggression and Racism

The activity of individual brain regions can be measured using functional magnetic resonance imaging. This makes it possible to display on slice images with millimeter resolution the brain areas that are engaged when a task is solved or emotions arise.

If test subjects were shown familiar faces from their own team, brain areas associated with positive emotions were activated; if they were shown unfamiliar faces, these brain areas were not active [45]. The brain functions underlying social identity and group aggression in supporters of rival sports teams were also studied with functional magnetic resonance imaging [46]. If one's own team lost or the opposing team won, then brain regions for negative emotions were activated; if one's own team won or the opposing team lost, then the brain's reward center (nucleus accumbens) lit up in this brain imaging method. The stronger the identification with the own group, the more pronounced these effects were in the tomographic brain images. The stronger the feeling of happiness associated with the victory of one's own team, the more likely aggressive action against the fans of the opposing team also became.

The limbic system of our brain, which is responsible for the emotional classification of what we perceive, reacts in a fraction of a second, almost reflexively, to the faces of people of a different skin color, even before we become aware of it. In several functional magnetic resonance imaging studies of black and white Americans, it was demonstrated that even below the temporal threshold for conscious perception, when faces of the other skin color are presented for only milliseconds, the amygdala becomes active [47–49]. This is a part of the phylogenetically ancient reptile brain and plays a central role in controlling fearful and aggressive emotions. The more pronounced racial prejudice, the more intensely the amygdala fired. Faces of the other skin color were also rated as more threatening than those of one's own. If a face of the other skin color was shown over a longer period of time so that it could be consciously processed in the neocortex, the activity of the amygdala was inhibited again in most individuals by these superior phylogenetically new brain cortical parts [48, 49].

These studies impressively suggest that primitive racist reflexes have always been firmly anchored in phylogenetically ancient brain structures in many, if not all, brains, but that they can be repressed by higher cognitive evaluations in higher-level neocortical cortical areas, provided that these higher brain regions are intact and have been filled with prosocial content in the course of education.

Conversely, one can conclude that the more primitively the brain reacts, the more pronounced racism is.

How strong an innate emotional division into "own" or "foreign" group is, can already be seen in children. Even at preschool age, they classify people according to race, have more negative views about those who do not belong to their own group, and find faces of another race angrier and more frightening than those of their own [50, 51].

Getting to Know Each Other Against Prejudice

The better people of different origins and races get to know each other and the more they share common goals, the more prejudice, xenophobia and racism disappear. This finding was confirmed by an extensive meta-analysis summarizing the results of 517 publications worldwide on this topic [52]. The more contact there was between groups, the more cooperative they became; and the effects were sustained, regardless of race, religion, skin color, national and cultural affiliation, or sexual orientation.

References

1. Möller-Leimkühler AM, Bogerts B (2013) Collective violence—neurobiological, psychosocial and sociological conditions. Nervenarzt 84(11):1345–1354, 1356–1358. https://doi.org/10.1007/s00115-013-3856-y
2. Wikipedia.org, Lorenz K. On Aggression. Wikipedia.org. https://en.wikipedia.org/wiki/On_Aggression. Published 2021. Accessed 26 Apr 2021
3. Thompson FJ, Marshall HH, Vitikainen EIK, Cant MA (2017) Causes and consequences of intergroup conflict in cooperative banded mongooses. Anim Behav 126:31–40. https://doi.org/10.1016/j.anbehav.2017.01.017
4. Goodall J (1971) In the shadow of man. William Collins Sons & Co., London
5. Wrangham RW, Glowacki L (2012) Intergroup aggression in chimpanzees and war in nomadic hunter-gatherers. Hum Nat 23(1):5–29. https://doi.org/10.1007/s12110-012-9132-1
6. Wrangham R, Peterson D (1996) Demonic males: apes and the origins of human violence. Houghton Mifflin Company, New York
7. Wrangham RW, Wilson ML (2006) Collective violence: comparisons between youths and chimpanzees. Ann N Y Acad Sci 1036(1):233–256. https://doi.org/10.1196/annals.1330.015
8. Lahr MM, Rivera F, Power RK et al (2016) Inter-group violence among early Holocene hunter-gatherers of West Turkana Kenya. Nature 529(7586):394–398. https://doi.org/10.1038/nature16477
9. Meller H, Shefzik M (2015) Krieg—Eine Archäologische Spurensuche. Begleitband Zur Sonderausstellung Im Landesmuseum Für Vorgeschichte Halle (War—an Archaeological Search for Traces. Accompanying Volume to the Special Exhibition in the State Museum of Prehistory Halle (Saale)). Halle (Saale): Landesamt für Denkmalpflege und Archäologie Sachsen-Anhalt. https://www.researchgate.net/publication/284142322_H_MellerM_Schefzik_Hrsg_Krieg_-_eine_archaologische_Spurensuche_Begleitband_zur_Sonderausstellung_im_Landesmuseum_fur_Vorgeschichte_Halle_Saale_6_November_2015_bis_22_Mai_2016_Halle_Saale_2015
10. Wikipedia.org. Tollense Valley Battlefield. Wikipedia.org. https://en.wikipedia.org/wiki/Tollense_valley_battlefield. Published 2021. Accessed 4 May 2021
11. Rummel R (1994) Death by government. Transaction Publishers, New Brunswick, NJ
12. Rummel R (1997) Statistics of democide: genocide and mass murder since 1900. Transaction Publishers, Rutgers University, New Brunswick, NJ
13. Gerlach C (2010) Extremely violent societies: mass violence in the twentieth-century-world. Cambridge University Press
14. Pinker S (2011) The better angels of our nature: why violence has declined. Viking Books Adult.
15. Pinker S (2011) The better angels of our nature: why violence has declined. Viking Books Adult, Preface
16. Harff B (2003) No lessons learned from the holocaust? assessing risks of genocide and political mass murder since 1955. Am Polit Sci Rev. 97(01):57–73. https://doi.org/10.1017/S0003055403000522
17. Rummel R (2020) Democide since World War II. https://www.hawaii.edu/powerkills/POSTWWII.HTM. Accessed 28 June 2020
18. Wikipedia.org. Indian Rebellion of 1857. Wikipedia.org. https://en.wikipedia.org/wiki/Indian_Rebellion_of_1857. Published 2021. Accessed 4 May 2021
19. Wikipedia.org. Anomie. Wikipedia.org. https://en.wikipedia.org/wiki/Anomie. Published 2021. Accessed 4 May 2021
20. Verdeja E (2016) Predicting genocide and mass killing. J Genocide Res 9(3):13–32. https://doi.org/10.1080/14623528.2020.1818478
21. Tajfel H, Turner J (1986) The social identity theory of intergroup behavior. Psychol Intergr Relations 5:7–24
22. Tajfel H (1982) Social psychology of intergroup relations. Annu Rev Psychol 33(1):1–39. https://doi.org/10.1146/annurev.ps.33.020182.000245

23. Tajfel H, Turner J (2001) An integrative theory of intergroup conflict. In: Hogg MA, Abrams D (eds) Intergroup relations. Psychology Press, New York, NY, pp 94–109
24. Böhm R, Rusch H, Baron J (2020) The psychology of intergroup conflict: a review of theories and measures. Researchgate.net. https://www.researchgate.net/publication/322666126_The_Psychology_of_Intergroup_Conflict_A_Review_of_Theories_and_Measures. Published 2018. Accessed 25 June 2020
25. Brown R (2000) Social identity theory: past achievements, current problems and future challenges. Eur J Soc Psychol 30(6):745–778. https://doi.org/10.1002/1099-0992(200011/12)30: 6%3c745::AID-EJSP24%3e3.0.CO;2-O
26. Densley J, Peterson J (2018) Group aggression. Curr Opin Psychol 19:43–48. https://doi.org/10.1016/j.copsyc.2017.03.031
27. Eckert R (2020) Radikalisierung in konflikttheoretischer Perspektive [Radicalization in a conflict-theoretical perspective]. In: Slama BB, Kemmesies U (eds) Handbuch Extremismusprävention, Gesamtgesellschaftlich Phänomenübergreifend [Handbook on the prevention of extremism, across all social phenomena], vol 1. Bundeskriminalamt, Wiesbaden, pp 213–269
28. Elias N, Scotson JL (1994) The established and the outsiders; a sociological enquiry into community problems, vol 2. Sage Publications Inc, London, Thousand Oaks, New Delhi
29. McKenzie J, Twose G (2015) Applications and extentions of realistic conflict theory: moral development and conflict prevention. In: Valsiner J (ed) Norms, groups, conflict, and social change; rediscovering muzafer sherif's psychology, vol 1. New Dost-Gözkan, Ayfer, Brunswick, NJ; Sonmez Keith, Doga, pp 307–325
30. Wikipedia.org. Realistic Conflict Theory. Wikipedia.org. https://en.wikipedia.org/wiki/Realistic_conflict_theory. Published 2020. Accessed 25 June 2020
31. Stephan WG, Ybarra O, Rios M (2009) Intergroup threat theory. In: Nelson TD (ed) Handbook of prejudice, stereotyping, and discrimination, vol 1. Psychology Press, Taylor & Francis Group, New York, pp 43–61
32. Obaidi M, Kunst JR, Kteily N, Thomsen L, Sidanius J (2018) Living under threat: Mutual threat perception drives anti-Muslim and anti-Western hostility in the age of terrorism. Eur J Soc Psychol 48:567–584
33. Obaidi M, Thomsen L, Bergh R (2018) "They think we are a threat to their culture": meta-cultural threat fuels willingness and endorsement of extremist violence against the cultural outgroup. Int J Conf Violence 12:1–13. https://doi.org/10.4119/UNIBI/ijcv.647
34. Zick A, Berghan W, Mokros N (2019) Gruppenbezogene Menschenfeindlichkeit in Deutschland 2002–2018/19 [Group-related misanthropy in Germany 2002–2018/19]. In: Schröter F (ed) Verlorene Mitte Feindselige Zustände, Rechtsextreme Einstellungen in Deutschland 2018/2019 [Lost Middle Hostile Conditions, Right-Wing Extremist Attitudes in Germany 2018/2019], vol 1. Friedrich-Ebert-Stiftung, Bonn, pp 53–114
35. Rees JH, Rees YPM, Hellmann JH, Zick A (2019) Climate of hate: similar correlates of far right electoral support and right-wing hate crimes in Germany. Front Psychol 10(2328):1–14. https://doi.org/10.3389/fpsyg.2019.02328
36. DW.com. Germany: Right-wing criminality at a record high. DW.com. https://www.dw.com/en/germany-right-wing-criminality-at-a-record-high/a-57421079. Accessed 7 May 2021
37. Popitz H (2017) Phenomena of power; authority, domination and violence. In Poggi G (ed). Columbia University Press
38. Schäfer-Vogel G (2007) Gewalttätige Jugendkulturen, Symptom der Erosion kommunikativer Strukturen, S.551[(Violent youth cultures—symptom of the erosion of communicative structures]. Band K 134. In: Albrecht H-J, Kaiser G (eds). Max-Plank-Institut für ausländisches und internationales Strafrecht, Berlin
39. Browning CR (1998) Ordinary men: reserve police battalion 101 and the final solution in Poland. HarperCollins, New York
40. Hoebel T, Knöbl W (2019) Gewalt Erklären! Plädoyer für eine entdeckende Prozesssoziologie [Explaining Violence! Plea For A Discovering Sociology of Processes]. Hamburger Edition HIS Verlagsgesellschaft. mbH, Hamburg

41. Neitzel S, Welzer H (2013) Soldiers: German POWs on fighting, killing, and dying. Reprint Ed. Vintage
42. Wikipedia.org. Crowd psychology. Wikipedia.org. https://en.wikipedia.org/wiki/Crowd_psychology. Published 2021. Accessed 6 May 2021
43. Wikipedia.org. The Crowd: A Study of the Popular Mind. Wikipedia.org. https://en.wikipedia.org/wiki/The_Crowd:_A_Study_of_the_Popular_Mind. Published 2021. Accessed 6 May 2021
44. Claus R (2018) Hooligans. Eine Welt zwischen Fußball, Gewalt und Politik [A world between football, violence and politics]. Verlag Die Werkstatt GmbH
45. Van BJJ, Packer DJ, Cunningham WA (2008) The neural substrates of in-group bias. Psychol Sci 19(11):1131–1139
46. Cikara M, Botvinick MM, Fiske ST (2011) Us versus them: social identity shapes neural responses to intergroup competition and harm. Psychol Sci 22(3):306–313. https://doi.org/10.1177/0956797610397667
47. Kubota JT, Banaji MR, Phelps EA (2012) The neuroscience of race. Nat Neurosci 15(7):940–948. https://doi.org/10.1038/nn.3136
48. Richeson JA, Baird AA, Gordon HL et al (2003) An fMRI investigation of the impact of interracial contact on executive function. Nat Neurosci 6(12):1323–1328. https://doi.org/10.1038/nn1156
49. Knutson KM, Mah L, Manly CF, Grafman J (2007) Neural correlates of automatic beliefs about gender and race. Hum Brain Mapp 28(10):915–930. https://doi.org/10.1002/hbm.20320
50. Baron AS, Banaji MR (2006) The Development Of Implicit Attitudes. Evidence of race evaluations from ages 6 and 10 and adulthood. Psychol Sci 17(1):53–58. https://doi.org/10.1111/j.1467-9280.2005.01664.x
51. Bigler RS, Jones LC, Lobliner DB (1997) Social Categorization and the formation of intergroup attitudes in children. Child Dev 68(3):530. https://doi.org/10.2307/1131676
52. Pettigrew TF, Tropp LR (2006) A meta-analytic test of intergroup contact theory. J Pers Soc Psychol 90(5):751–783. https://doi.org/10.1037/0022-3514.90.5.751

Chapter 19
Sexual Violence

In the vast majority of cases sexual violence is committed by men. What are the motives? Who are the perpetrators?

Definition

The International Criminal Court has defined all forms of sexual violence, particularly rape, sexual slavery, enforced prostitution, forced pregnancy, enforced sterilization or any other form of violence of comparable gravity as crimes against humanity [1]. The most common forms of sexual violence are rape within marriage or dating relationships, or by strangers, systematic rape during armed conflicts, sexual abuse of children or disabled people, and forced prostitution and trafficking of young women for the purpose of sexual exploitation. Victims are not only women but also—albeit to a much lesser extent—men, especially in prisons [2].

The term "sexual violence" has been increasingly replaced by "sexualized violence" in recent years, as it has been argued that it is less a sexually motivated crime than an act to express power and dominance over the victim. The common motive would be the experience of one's own assertiveness, rather than sexual pleasure, in order to compensate for insults and self-esteem deficits.

This view can be questioned with the argument that obtaining sex is the strongest motive for male sexual violence and that male sexuality primarily serves not the subjugation of women but sexual pleasure. Men usually fight for power against male rivals and enemies, not against women.

The controversy over the terms "sexual" and "sexualized" violence can probably be resolved with regard to the brain-biological correlates of aggression and sexuality. The neuronal centers for appetitive-hedonistic aggression and sexual behavior are located next to each other in the brainstem (hypothalamus, see Chap. 16 and Figs. 6.4 and 13.2) and influence each other in a mutually inhibitory or facilitatory manner depending on the situation at hand. The location of the two centers in close proximity and their synergistic or antagonistic interaction in the deep layers of our reptilian brain explains why both violence and sexuality are not only associated with archaic primal instincts, but are also accompanied by the strongest emotional and vegetative arousal. Hedonistic violence and sexual acts have a certain affinity in some people, as they can stimulate their reward center with both archaic components of our mentality simultaneously. If a sexual motivation is added to perpetrators who already have a predisposition to hedonistic violence, the risk of sexual force is particularly high.

Frequency

A recent WHO report based on data from the period 2000–2018 revealed that the prevalence of violence against women is a global public health problem of pandemic proportions affecting hundreds of millions of women [3]. According to this report, the global estimates for the combined prevalence of intimate partner and non-partner sexual violence were 30 percent for women aged 15 years and older.

Determining the frequency of sexual violence is fraught with the problem of high underreporting rates. Many victims do not report such acts, which mainly occur in their social surroundings. In representative surveys conducted in Europe in 2014 and in the US in 2015, between 1 and 2% of women reported being a victim of rape in the 12 months prior to the survey. In the European survey, 5% of women, and in the US survey, four times as many (20%) women, reported having experienced rape at least once in their lifetime [4, 5]. According to a US survey, no more than 16% of cases were reported to police [6]. One in 14 men (=7%) also reported having been the victim of a sexually violent act in their lifetime [5].

A large part of sexual violence is directed against children and adolescents. Child sexual abuse is prevalent in all social classes in both low- and high-income countries.

According to UNICEF reports (2020), [7, 8] worldwide at least 120 million girls—about 1 in 10—have been forced to engage in sexual acts or have been otherwise sexually exploited. Five percent of girls aged 15–19 (around 13 million) have experienced forced sex during their lifetime. The actual figure should be even higher because of the high number of unreported cases. In most instances, victims and perpetrators know each other.

Around 90% of perpetrators are males. Girls typically report rates of victimization two to three times higher than boys. Adolescent girls in the US and in Spain reported victimization rates of 13–15% during the previous year.

One in 20 men admitted online sexualized behavior towards children known to be below the age of 12 [8].

The situation has deteriorated over the last decade since the Internet has opened up a rapidly growing global market for the consumption of child sexual abuse materials.

Types of Perpetrators: Risk Factors

98% of perpetrators of sexual violence are men. The social and individual risk factors for male sexual aggression are manifold and range from tolerance by the respective society, personality disorders and sexual deviance to alcohol and drug use [9, 10].

Forensic psychiatrists have sought to identify characteristics of perpetrators of sexual violence. A variety of offender types have been described:

- Sexually deviant offenders (sado-masochists)
- Women-hating "machos"
- Affective highly charged conflicts from relationship break-ups
- Dissocial perpetrators
- Late pubertal juvenile offenders
- Moronic perpetrators
- Assertive irritable offenders
- Socially disintegrated chauvinistic offenders
- Psychopathologically highly conspicuous offenders.

Although the very diverse typology of sexually violent offenders militates against a uniform picture, the group of dissocial, psychopathologically conspicuous and intelligence-diminished perpetrators is particularly frequently represented in all classifications. The heterogeneity of the forms of sexual violence and of the perpetrators does not allow a one-dimensional answer to the question of causes. Different types of perpetrators use different forms and intensities of sexual violence under different circumstances. The question of whether sexual violence is more of a violent or a sexual offense is therefore probably misplaced in its exclusivity. Both components come together in the described perpetrator types in predisposing situations.

Other more general risk factors mentioned are: [11].

- Alcohol and drug abuse
- Lack of empathy
- Coercive sexual fantasies
- Hostility towards women
- Hypermasculinity
- Adherence to traditional gender role norms
- Delinquent peers.

The set of conditions for the occurrence of sexual violence can also be explained with the help of the general aggression model (GAM, see Chap. 12, Fig. 12.5) just as well as the conditions for most other forms of violence. The interplay of personality constitution, cultural and social background, current stimulating and inhibiting situational factors, and the individual's current state of mind determine whether or not a sexual assault occurs. What is special about sexual violence is that when it is carried out, sexual stimulation and hedonistic violence combine to serve the reward system in the perpetrator's brain.

War and Sexual Violence

Sexual violence in peacetime can be limited to certain types of perpetrators or exceptional circumstances. The situation is different in war. Here, completely normal men are capable not only of mass killings, but also of mass rapes. The principle of male warlike behavior has always been the same in its brutal variant since the beginning of warlike conflicts in the history of mankind: the men of the enemies are killed, the women raped, enslaved or also killed. On the part of the victors, it is not only about humiliating the women of the defeated, but also about further increasing the euphoric feeling of victory through sexual pleasure by rape. In analogy to mass killings in war, one can postulate a shift in the frame of reference [12] for groups of soldiers in such situations as well, away from the moral norms commonly applicable in bourgeois life toward common warlike atrocities (see Chap. 18).

Examples on a drastic scale from recent history include the behavior of German soldiers at the beginning and soldiers of the Soviet army at the end of the Second World War; the orgies of rape by Japanese military in conquered Asian territories in the 1930s and 1940s; and the incidents during the genocides in Bangladesh in 1971–1972, in the Bosnian war in 1993 and in Rwanda in 1994.

For Germany, in the immediate post-war period the number of women who were raped is estimated at two million, mostly by Soviet soldiers before the military leadership, concerned about the discipline of the troops, took action against them.

Phylogenetic Aspects

The genes of rapists are also passed on and with them the genetic basis of their mentality—one reason why this phenomenon has always been present in the history of mankind.

The behavior of our ancestors regarding sexual violence in war might not have been very different from that of their descendants. An interesting clue to this could be the comparison of genes from archaeologically recovered remains of Neanderthal man with our genes. The male Y sex chromosome of Neanderthal man, who lived for several thousand years in Europe at the same time as *Homo sapiens* and died out 30,000 years ago, can never be detected in modern humans, but the X chromosome, which is present twice in females but only once in males, like the Y chromosome, can be found in modern men [13]. If our primitive ancestors behaved in the same way as their descendants and during ethnic conflicts with Neanderthal man eliminated his male fighters and raped the women of the defeated, only the female X chromosome, but not the Y chromosome of the Neanderthal men had the chance to be inherited further. The Y chromosome of the *Homo sapiens* victors, on the other hand, which helped shape their mentality, continued to exist in their male descendants and formed the genetic basis for their future behavior in war right up to the present day.

References

1. Elements of Crime—Article 7 Crimes against humanity. International Criminal Court. https://www.icc-cpi.int/resourcelibrary/official-journal/elements-of-crimes.aspx. Published 2011. Accessed April 30, 2021
2. Wikipedia.org. Sexual violence. https://en.wikipedia.org/wiki/Sexual_violence. Accessed April 30, 2021
3. WHO (2021) Violence against women prevalence estimates, 2018. https://www.who.int/publications/i/item/violence-against-women-prevalence-estimates
4. FRA European Union agency for fundamental rights. Violence against women: an EU-wide survey. fra.europa.eu. https://fra.europa.eu/sites/default/files/fra-2014-vaw-survey-factsheet_en.pdf. Published 2012. Accessed July 13, 2020
5. Smith SG, Zhang X, Basile KC et al National intimate partner and sexual violence survey: 2015 data brief-update release. National Center for Injury Prevention, Atlanta, GA. https://www.cdc.gov/violenceprevention/pdf/2015data-brief508.pdf. Published 2015. Accessed July 13, 2020
6. Kilpatrick DG, Resnick H, Ruggiero K, Conoscenti L, McCauly J Drug-facilitated, incapacitated, and forcible rape: a national study. https://www.researchgate.net/profile/Dean_Kilpatrick/publication/224918954_National_Prevalence_of_Posttraumatic_Stress_Disorder_Among_Sexually_Revictimized_Adolescent_College_and_Adult_Household-Residing_Women/links/5891f0f8aca272f9a5582ebe/National-Preva. Published 2007. Accessed July 13, 2020
7. UNICEF. Sexual violence against children. https://www.unicef.org/protection/sexual-violence-against-children. Accessed April 30, 2021
8. UNICEF (2020) Action to end child sexual abuse and exploitation: a review of the evidence. https://www.unicef.org/media/89096/file/CSAE-Report-v2.pdf. Published 2020. Accessed April 30, 2021
9. Krahé B (2018) Violence against women. Curr Opin Psychol 19:6–10. https://doi.org/10.1016/j.copsyc.2017.03.017
10. Krahé B (2013) The social psychology of aggression, vol 2. Psychology Press, Taylor & Francis Group, London
11. Violence prevention—sexual violence—risk and protective factors. Centers for Disease Control and Prevention. https://www.cdc.gov/violenceprevention/sexualviolence/riskprotectivefactors.html. Published 2021. Accessed April 30, 2021
12. Neitzel S, Wetzler H (2017) Soldaten - Protokolle Vom Kämpfen, Töten Und Sterben. Fischer-Verlag, Frankfurt a.M.
13. Mendez FL, Poznik GD, Castellano S, Bustamante CD (2016) The divergence of neandertal and modern human Y chromosomes. Am J Hum Genet 98(4):728–734. https://doi.org/10.1016/j.ajhg.2016.02.023

Chapter 20
Religion and Violence

At the heart of the teachings of the great religions is the promise of a paradisiacal afterlife for those who follow the right path in compliance with divine commandments and the teachings of the great founders of religions. On earth, however, this path often proves to be quite arduous and is not always characterized by peacefulness. The right path to goodness has been accompanied by violent collateral damage in all religions and cultures throughout history right up to the present day. Why is there violence in the name of religion?

Common Characteristics of the Great Religions

Religious faith and belonging to a religious community are sense-giving and solidifying for all religions worldwide. Common faith, common traditions, common rituals, ceremonies and services promote a sense of community and social identity, a feeling of belonging and security. Shared religious values, ethical and moral concepts form a stable basis for mutual reliability and joint action, and provide consolation in times of crisis. The high acceptance and success of the five most widespread world religions—Christianity, Islam, Judaism, Hinduism and Buddhism—as well as the Far Eastern teachings of Confucius and Jnatraputra (both sixth century BCE), can be explained by the fact that they declare prosocial, compassionate behavior (often, however, only toward members of one's own religious community) to be the core of religious duties in order to achieve a more bearable life in this world and reward in the hereafter.

Typical for the great world religions is a spirit–matter dualism, which offers explanations for phenomena that cannot be reasoned with earthly occurrences and the experiences of the believers by invoking supernatural powers (God, angels, Satan). This satisfies a need for causal closure and teleological (i.e., meaning and purpose-seeking) thinking. The course of the world as well as the meaning of one's own existence thus appears more plausible.

Violence in the Name of the Religions

Despite the teachings of all religions requiring patient, peaceful and compassionate behavior followed by the vast majority of believers, a not inconsiderable number of religious adherents in the course of history have committed massive acts of violence, especially against people of other faiths and apostates. Atrocities and the killing of non-believers, pagans or heretics could be justified on religious grounds. It was hardly ever the contents of faith itself that provided the reason for murder. Rather, the motives were those that usually underlie proactive violence, dressed up in religious terms: obsession with power, feelings of superiority, striving for dominance, greed and fanaticism. Often—as, for example, in the case of the medieval Inquisition or witch burnings—hedonistic violence can also be assumed as the central motive: satisfaction through torture or the act of killing in itself, for the justification of which religion was misused.

Islam

Since the activities of Islamist jihadist terrorist groups, especially the so-called Islamic State (IS), which began with the attack on the World Trade Center on September 11, 2001, it has been widely believed that Islam is a particularly violent religion. However, in its original self-understanding the entirety of the Muslim community, the *ummah*, emphasized virtues such as compliance, patience and mercy by caring for the needy, freeing slaves and commanding people to be helpful and kind to others. When provoked, Muslims of the *ummah* should not retaliate, but leave revenge to Allah [1]. However, the Quran's statements regarding the use of military force are not consistent. Sometimes Allah demands restraint instead of fighting, sometimes he calls for war. After the Prophet Mohamed's return to Mecca (630 CE), Allah commanded war against non-Muslims wherever and whenever they were encountered. Also, in considering the connection between violence and Islam, it should be noted that the prophet Mohamed was not only the founder of the religion, but also a military strategist and commander. Islam owes its early successes and rapid spread not only to the prosocial principles of the *ummah*, but also to the raids, attacks and battles commanded by Mohamed, which he led victoriously against his opponents in Mecca after the *hijra* in the years 622–630 CE. On the other hand, there are Quranic verses that call not to retaliate but to forgive [1].

Since then, militant, jihadist Muslim groups have used the "Sura of the Sword" as a justification for fighting unbelievers: "...kill the idolaters wherever you find them, seize them, besiege them and waylay them from every ambush! But if they repent, perform the prayer and pay the tribute, let them go on their way! Surely Allah is Oft-Forgiving and Most Merciful." (quoted after [1]). Throughout the Quran there is thus a contrast between incitement to fight and mercy towards those who ask for peace. The prohibition on killing contained in the Ten Commandments valid for

Christians and Jews is relativized in the corresponding *sura*s of the Quran: "Do not kill human life, which God has declared inviolable, unless justified. If someone is killed unjustly, we give his closest relatives the authority to avenge him" (Sura 17, 33) [2]. Elsewhere it is said (Sura 5, 31 and 32): "... if someone kills a person without that person having committed murder, or without any mischief having occurred in the country, it shall be as if he had killed all mankind; and if someone preserves the life of a person, it shall be as if he had preserved the life of all mankind." And (Sura 5:32): "... The measures for those who wage war against Allah and His Messenger and seek to cause mischief in the land shall be that they may be killed or crucified, or that their hands and feet may be cut off alternately, or that they may be driven out of the land...". Other *sura*s of the Quran contain content similar to the biblical Ten Commandments [3].

The term jihad means holy religious war, which is justified by a perceived threat to Islam, especially from the Western world, and which, according to extremist Salafists, is to be waged as a kind of self-defense, even armed. Anyone who falls in battle against infidels is revered as a martyr, just like a suicide bomber, who moves directly into paradise and is showered with the finest rewards in the afterlife. The glorification of jihadist violence also includes fight songs that are disseminated on the Internet and video portals to recruit young people to fight against infidels and risk a martyr's death on the way to paradise [4].

The most militant branch of Islam is Salafism, but this is not always synonymous with extremism or even terrorism. Only a smaller proportion of Salafists come under the description of jihadist Salafism. Salafists see themselves as the true Muslims, referring to the founding generation of Islam. Not only members of other religions or atheists are regarded by Salafists as *kuffar* (infidels), but Muslims who do not practice their faith as the extremist ideology would have them do are also declared infidels. Such fundamentalism within Salafism is particularly pronounced in Wahhabism, which emerged in Saudi Arabia in the eighteenth century. Violence-oriented Salafist groups invoke the "sword *sura*" of the Quran and call for the killing of *kūffār* [5].

The eternal battles in the Gulf region between Shiites, Sunnis and Wahhabis, with Iran, Iraq and Saudi Arabia as the main actors, are not only power-political but also religious disputes between different denominations of Islam and are reminiscent of the conditions in Europe during the Thirty Years' War.

Most of the victims of the attacks by jihadist terrorists in the Middle East and the Gulf region are other Muslims.

Christianity

The statements of the New Testament of the Bible on the handling of violence are also not unambiguous. On the one hand, in the Gospel of Matthew, Jesus' Sermon on the Mount (Matthew 5–7) says: "Blessed are the meek, for they shall inherit the land....—Blessed are the merciful, for they shall find mercy...—Blessed are the peacemakers, for they shall be called children of God... ...Love your enemies and pray for those who persecute you." On the other hand, when Jesus sends his disciples (Matthew 10:34–36), he also says: "Do not think that I have come to bring peace on earth! I have not come to bring peace, but the sword; for I have come to divide the son with his father, and the daughter with her mother, and the daughter-in-law with her mother-in-law."

Regarding the early history of Christianity, it should be remembered that it owes its spread in the Roman Empire to the Battle of the Milvian Bridge near Rome, won by the Emperor Constantine in the year 312 CE. According to legend, a cross appeared to Constantine before the battle bearing the words "in this sign you will be victorious." In any case, the emperor attributed his victory to the assistance of the Christian God, and therefore ceased the persecution of Christians practiced by his predecessors and favored their religion. His successor, Emperor Theodosius, elevated Christianity to the status of state religion in 380 CE, with the result that from then on non-Christians were harassed and persecuted. Theodosius enacted laws with the threat of punishment against paganism, allowed Christian monks to raid pagan villages and shrines, and expelled or killed pagan priests [6]. This was the first violent Christianization. The Christians, who had earlier faced intense persecution even in the Roman Empire, now turned the tables on the pagans.

Christian violence in the Middle Ages was directed against all non-Christians, especially Jews, Muslims and heretics. In each of the Crusades that took place between 1095 and 1208, Christians first attacked Jews in their own country. In the capture of Jerusalem in 1099, the Christian conquerors, in a bloodlust, massacred all the inhabitants they could find, mostly Muslims; the dead are said to have numbered 30,000. Afterwards, the Crusaders marched to the Church of the Holy Sepulchre singing hymns and celebrating the Easter liturgy [7]. One of the most influential Crusade preachers was the Cistercian monk Bernard of Clairvaux (+1153). He preached that a Christian should rejoice when he sees the pagans scattered and driven out. To die fighting for the Lord was a high merit. He praised the Knights Templar, an order of knights made up of militarily equipped and organized monks whose goals included killing the enemies of Christ. Bernard of Clairvaux was canonized by the Church and has remained a frequently chosen patron saint to this day [8].

The Crusades were motivated not only by religion but also by power politics. Pope Urban II drove the knights of Christendom to embark on them in order to expand the power of the Church and create a papal monarchy dominating Christian Europe. Those who fell fighting Muslims were absolved of all sins by papal decrees. The Templars were a militaristic order devoted to the Pope. They themselves were wiped out 200 years after the last Crusade. As a fighting force subordinate to the Pope, the Templars seemed too dangerous to the French king. The latter confiscated their goods, threw the members of the order into dungeons, and had them tortured and burned at the stake [9].

The Crusaders behaved with brutality similar to that of the conquest of the Holy Land in the Albigensian Wars in France (1209–1229). The Albigensians were a socio-religious movement that deviated from Church doctrine, believing in a good God and an evil God; they despised the traditional Christian Church with its accumulation of vast material wealth. They were declared heretics at the behest of the Pope, imprisoned, tortured and burned; their cities were destroyed until their complete extinction.

Those who questioned papal teaching were persecuted and eliminated for centuries. Thousands of apostates, declared heretics, fell victim to the Inquisition, carried out mainly in Spain, in the name of the Church in the sixteenth and seventeenth centuries.

After the discovery of America, the Christian conquerors plundered and massively decimated the indigenous population. Pope Alexander VI allowed the Spanish royal family to wage a "just war" against all indigenous peoples who resisted the European colonizers. According to one estimate, between 1519 and 1595, the population of Central Mexico dropped from 17 to 1 million people, and the Inca population was halved [10].

Even among themselves, the Christian religions based on the Holy Scriptures did not always behave peacefully. The Huguenots, who emerged from the Reformation movement, came into conflict with Church doctrine and the Catholic French royal family, leading to the Huguenot Wars (1562–1594). The climax was the St. Bartholomew's Day massacre of 1572. Huguenots from all over the country had come to Paris for what was meant to be a reconciliation. At the instigation of the Catholic royal house, all the Huguenots who could be found in Paris were slaughtered that night and an extermination process was initiated throughout the country to which tens of thousands fell victim [11]. When Pope Gregory XIII received the news of this, he is said to have intoned a *Te Deum* in the Vatican. The surviving Huguenots fled to neighboring European countries.

The greatest European inferno, triggered by religious conflicts between Catholics and Protestants but also by power-political ambitions of European rulers, was the Thirty Years' War (1618–1648). The fierce battles between the adherents of these two great Christian religions and the resulting devastation of entire regions claimed five to six million lives, about one-third of the population of Central Europe at the time.

Another example of bloody massacres of Christians against Christians is the conquest of the Irish town of Drogheda in September 1649 by Cromwell's troops. The Puritan conquerors wanted to eliminate the Catholic Irish or reduce them to slave status, and they largely succeeded. In doing so, Cromwell saw divine will on his side and was certain that through his actions he was accomplishing God's work in history by suppressing and destroying the infidels [12]. The conflict in Northern Ireland, which can be traced back to these historical contexts, continues to this day.

With the Enlightenment in the early eighteenth century and the increasing importance of humanism, the violent side of Christianity gradually faded away. Instead of maintaining their authority by means of violence, exclusion or threats of punishment in the afterlife, the churches today see their central task in spiritual, charitable and pastoral work. This was also their most important mission in the past, although at times it escalated into violence. Christian as well as other religions are today a significant component of society for the satisfaction of religious needs and the promotion of compassion and humanity.

Judaism

According to the Old Testament, after the exodus of the Israelites from Egyptian captivity on Mount Sinai, Moses received from God the Ten Commandments; among them the commandment "Thou shalt not kill." The events immediately following this are recorded in the Old Testament thus (Exodus 32:26–29):

> … and Moses stood in the gate of the camp (of the Israelites), and said, Come unto me, that is for Jehovah. And there gathered unto him all the sons of Levi. And he said unto them: Thus saith Jehovah, the God of Israel; Put every man his sword upon his thigh, and go from gate to gate in the camp, and slay every man his brother, and every man his friend, and every man his neighbour. And the sons of Levi did according to the word of Moses: and there fell of the people that day about three thousand men.

All who were not for Jehovah were killed. On entering the Promised Land, Moses instructed his people to wipe out the indigenous population there: "The Lord your God will deliver them up to you. You shall execute his judgement on them and kill them…" [13]. Further, the Old Testament reports (Deuteronomy 2:3), "At that time we conquered all the cities. We consecrated the whole male population, the women, the children and the old men to destruction; we left none alive" and (Deuteronomy 3:3–3:7): "…the Lord our God also gave Og king of Bashan and all his people into our power. We smote him and left none alive. … We consecrated the whole male population and the women, children and old men to annihilation. All the cattle and what we had plundered in the cities we kept as booty." According to the biblical account, when the Israelites came upon the city of Jericho, located in the Promised Land, the walls of the city collapsed with the help of Jave; after that all the inhabitants of Jericho were killed, 12,000 men and women, and the city was made a ruin [13].

In the Hebrew Bible, too, there are many juxtapositions of God's approval of violence with calls for compassionate behavior. In Judah, the prophet Isaiah spoke out against the exploitation of the poor, the dispossession of peasants and the use of violence, and for the rights of widows and orphans. Another passage of the Old Testament states (Exodus 21:23–25): "An eye for an eye, a tooth for a tooth," by which was meant the retaliation of like with like and not with even worse. On the other hand, a commandment of the Torah of Judaism, which is considered the word of God (Lev 19:18 EU), says: "You shall not take vengeance on the children of your people, nor bear a grudge against them. You shall love your neighbour as yourself." In addition, the testaments of the 12 patriarchs, written around 200 BCE, demand: "Love the Lord in all your life and one another with a true heart" and "… love the Lord and your neighbour, have mercy on the weak and the poor" [14].

The Bible and the Quran are similar in that passages that call for violence and justify it are difficult to reconcile with passages that condemn violence and call for humanity.

Throughout history, from antiquity to the Middle Ages and into the twentieth century, Judaism itself was subjected to the most intense exclusion and persecution. This is probably also one reason why, for Orthodox Jews, Israel has remained to this day the land promised by God only to his chosen people. The violent claim of others to this land is met with counter-violence and armament. Both the Shiite Hezbollah (which translates as Party of God) fighting the State of Israel and the Jewish settlers in the West Bank believe that God is on their side.

Buddhism and Hinduism

At the heart of the Buddha's teachings (fifth century BCE), which aim to free life from pain and adversity, is non-violence, rejection of hatred and greed, substituted by friendship, affection and commitment to all people. Buddhism is therefore often regarded as the most peaceful of all religions.

The rules of life of the various Hindu traditions that emerged as early as the second millennium BCE also include non-violence, suppression of anger, greed and slander, compassion for creatures, mildness and modesty. Neither fellow human beings nor other living beings should be harmed, which is one reason why many Hindus are vegetarians.

Buddhism and Hinduism also have their dark sides. In Myanmar, under the influence of Buddhist monks' agitation against Muslims, hundreds of thousands of Rohingya Muslims were displaced to neighboring Bangladesh in 2017, thousands were raped or murdered, villages were burned and razed to the ground. In other parts of the country, Koran schools were closed under pressure from radical Buddhists, and Buddhist mobs indiscriminately attacked Muslims from the neighborhood [15].

Far from being an exception, the events in Myanmar are part of a long history of violence perpetrated by Buddhists. Leading Japanese Zen Buddhists cheered on the Japanese war of aggression against China that began in 1937, in which the Japanese armies unrestrainedly looted, murdered, raped and burned; [16] 15 million Chinese were killed. In the capture of the city of Nanking alone, several hundred thousand Chinese civilians were massacred by Japanese soldiers. In Sri Lanka, the decades-long civil war between Buddhist Sinhalese and Hindu Tamils, legitimized by Buddhist monks, came to an end in 2009.

Hindu terror against people of other faiths is also not uncommon in recent history. The demolition of the Babri Mosque in India in 1992 by Hindus to humiliate the Muslim minority caused an international scandal. This was followed in retaliation by an attack by Muslims on a train carrying Hindu pilgrims, which in turn was avenged by Hindus with a massacre of Muslims [17]. Hindu fundamentalists carried out a brutal persecution of Christians in the Indian state of Odisha in 2007–2008. In a pogrom-like action, more than 300 villages and almost 5,000 houses and churches were destroyed; many Christians lost their lives [18].

On the other hand, there was massive persecution of Hindus. From 1946 to 1970, at least four million Hindus fled from the East Bengal province to escape Muslim invaders from West Pakistan. Tens of thousands of Hindus were killed in repeated pogroms. However, the violence did not only come from one side in the course of the conflict. After the Pakistani military was driven out with the help of the Indian army, West Pakistani prisoners were killed in a revenge action; thousands of people of non-Bengali origin were killed by Hindus. Finally, the violence in this conflict, both religious and ethnic, was multipolar: Muslims against Hindus, Hindus against Muslims, Hindus and Muslims against Christians, Bengalis against non-Bengalis. It was the Hindus who suffered the most from this [19].

Honor killings, which were widespread in the past, still occur in some circles of archaic Hindu and Muslim societies governed by tribal traditions. The most common victims are young women who refuse to enter into an arranged marriage or who want to divorce from their husband, as well as those who refuse to follow orthodox religious rules.

Sects

Extreme violence has been perpetrated not only by members of the major world religions but also by members of sects. The following are two particularly striking examples.

In November 1978, a murderous action by the Peoples Temple Christian sect in Guyana led by the American preacher Jim Jones caused worldwide shock. Because of reports of abuses in the sect, concerned relatives of sect members, accompanied by some sect opponents and a US congressman, visited the Jonestown settlement founded by Jones. Some breakaway members important to the cult wanted to leave the cult on that occasion. Assuming that Jonestown would now be overwhelmed by the evil forces of the world, Jones and followers attacked the group with weapons on the airfield as they were leaving. Five people died and 11 were injured. The following night, 911 of the Jonestown residents killed themselves or each other by ingesting or injecting the poison cyanide, either voluntarily or under duress; those who tried to escape were shot dead [20].

In March 1995, during the morning rush hour, the end-times preacher and leader of the Aum Shinrikyo sect and followers poured out the highly potent nerve agent sarin into the Tokyo underground. Several thousand people were injured, 11 people died instantly. The sect proclaimed the imminent end of the world, but felt it was not taken seriously and therefore used the attack to make its point [21].

In view of the immense acts of violence committed in the name of religion, one might assume that throughout history persecution, expulsion, torture, murder and war were predominantly carried out by religious representatives and their followers. But this is not the case. Secular regimes carried out massacres on a scale far exceeding that of religious conflicts. All of the ten genocides and pogroms of the twentieth century listed in Table 18.1 (Chap. 18) with more than one million deaths each—a total of more than 200 million deaths—were instigated by secular dictators who did not have much in common with religion or even fought against it.

Nevertheless, an explanation is needed as to why, throughout history and in all confessions, killing was so often seen as a sacred act to be performed in the name of God.

Psychological and Sociological Explanations

Throughout history, violence has been practiced indiscriminately by all religions in all parts of the world and by all ethnic groups, even if the intensity of the violence varied between different time periods and phases of non-violence, alternating with religious wars and persecutions of heretics. The omnipresence of religiously motivated or justified violence suggests that it belongs to the universals of human nature, which are inherent in the human psyche because they have evolved during the phylogenesis of humankind—like individual and collective violence in general—as a negative mortgage of our behavioral repertoire.

The general psychological and sociological principles for the occurrence of violence presented in the previous chapters also apply to violence by religious representatives or communities. The principle of social identity (see Chap. 18, "Collective Violence"), belief in the superiority of one's own (religious) group and devaluation of another, also applies to religious conflicts. Feelings of threat and the resulting conflicts are common to both religious and secular conflicts. The decisive factor is not the respective religious content, but simply the fact that two different faith groups set apart from each other clash; the phylogenetically shaped preconditions for collective aggression are just as present as in conflicts without a religious background. These are cohesion and identification with one's own group with a hostile attitude towards the foreign group, often paired with a hedonistic motivation for group violence. Religious justification, as with the Crusaders or Islamist fundamentalists, according to the motto, "kill the infidels, then the reward in the hereafter is certain", increases the desire to attack and accelerates the killing by eliminating any scruples that may exist. In addition, the hoped-for victory over the others is supposed to raise self-esteem.

Alleged orders from God to kill originate in the brains of those who want to murder and can be classified as products of the kinds of brain-biological activities that underlie proactive and hedonistic violence; they are far from being of supernatural origin.

In the torture and killing of heretics by the medieval Inquisition and in the witch burnings, the perpetrators invoked the need to eliminate evil from the world. It is reasonable to assume that religious arguments were used in the run-up to the brutal torture and execution methods applied in these cases in order to be able to act out hedonistic feelings and sadism.

Especially for monotheistic religions, an authoritarian insistence on the absolute truth of beliefs with punishment of those who question religious dogmas and do not want to follow them has been typical throughout history. This corresponds to the universal nature of those who possess power and claim authority for themselves; it is thus based on a general psychological or sociological principle of maintaining power. The claim to possess and maintain power goes hand in hand with the claim to possess the truth.

The claim to power and influence by religious administrators is greatest when as many as possible follow the teachings they propagate. If the teachings are doubted, the ideological foundation of the power base is attacked, which is why apostates, unbelievers and heretics must be eliminated as quickly as possible. The same applies to ideologically based dictatorial secular regimes.

The demand for absolute authority and obedience from their subordinates was underpinned by more than a few secular rulers declaring themselves to be godlike beings, from the Egyptian pharaohs to the Sun Kings of the Incas, and to the Japanese emperor, who was only deprived of this status at the end of the Second World War. Numerous European rulers, among them the German Emperor, invoked a divine nature, which meant that they regarded their office and actions as divinely legitimized. This did not stop them from instigating wars.

Neuroscientific Explanatory Models for the Relationship Between Religion and Violence

Religion is as old as humanity itself, as evidenced by the ring sanctuaries found in many parts of Europe, which were built over a period of several millennia BCE. Religious rituals can be found in all primitive cultures and promote group cohesion and like-minded behavior there as well as in religious communities of the modern world. The social value of religion in the course of human evolution consists of cooperation, mutual trust based on the same beliefs, and common action based on the same values and goals. This offered survival advantages over groups that did not have these properties. Therefore, the phylogenesis of *Homo sapiens* favored brains in which a predisposition to religious feeling and behavior was present. From an evolutionary (and thus scientific) point of view, this is the reason why religion is still demanded by many human brains today and will continue to be in the future.

Just like aggression and violence, religious feelings and actions are also results of processes in the human brain. This statement seems to be incompatible with the self-understanding of most religions, according to which supernatural forces or God himself have an influence on religious thoughts and actions. Above all, the major monotheistic (believing in only one God) world religions are based on a dualistic world view. According to this, spirit and soul on the one hand, body and natural things on the other, belong to separate worlds, even if they influence each other. The philosophical discussion about this is far from over. However, recent neuroscientific research has shown that religious feelings, like the tendency to violence, are linked to the function of overlapping circuits in the brain, at the center of which are the limbic structures of the middle temporal brain and areas of the frontal brain. [22, 23]. Disturbances in these brain areas can cause abnormal religious phenomena in some brain diseases [24]. The same circuits are also involved in the development of aggressive as well as prosocial behavior. Aggression, compassion and religious feelings thus have partly overlapping brain-physiological correlates.

Some brain researchers argue that religious belief is a product of our brain and thus a brain-physiological and therefore purely worldly matter [23]. This is supported by several functional nuclear magnetic resonance studies that have investigated the connection between brain physiology and religious feelings in recent years. When faith content was presented, believers mainly showed activation of areas of the lower and middle frontal brain and the temporal brain [22, 23] These are brain areas that also play a significant role in controlling aggressive behavior (see Chaps. 6 and 7, Figs. 6.8 and 7.1).

In devout Mormons, brain activity during spiritual devotion was studied by functional tomography. An activation of the nucleus accumbens (see Chaps. 6 and 13; Figs. 6.4, 6.8, and 13.2), a central switch point of the reward system of our brain, was detected [25]. Everything that activates the reward system, including religious ecstasy, satisfaction of elementary drives, social success—but also hedonistic violence—is perceived as pleasurable and is sought again when successful (see Chap. 13).

There is thus a remarkable overlap of the neuronal networks that are responsible for controlling violent behavior and also become active in religious feelings and empathy. This may provide a neuroscientific explanation for the close proximity of religion and fellow humanity on the one hand and violence on the other, and thus of primal human characteristics whose neuronal correlates can become active with or against each other in the same functional brain systems.

Religious Phenomena and Violence in Brain Diseases

The close connection between brain function, religion and violence is also expressed in the symptomatology of some brain diseases. Epilepsies originating in the limbic structures of the temporal brain (amygdala, hippocampus and neighboring structures (see Chap. 6, Fig. 6.5), do not manifest themselves in the form of classic epileptic seizures with sudden falling, trembling, cramping of the muscles and unconsciousness, but in states of peculiar, altered-consciousness behavior, often associated with strange movements and sensory phenomena that resemble hallucinations and are sometimes interpreted by the affected person as supernatural experiences. Up to a quarter of temporal lobe epileptics with pathological changes in the brain tissue in the region of the amygdala have massive aggressive outbursts in the course of the seizure event [26] (see Chap. 6). A small proportion of such patients (about 4 percent) [27] exhibit personality traits in everyday life between seizures that are characterized by hyper-religious behavior and writing mania (polygraphy). Such a diagnosis is said to have applied to the apostle Paul (+65 CE) [24]. After having previously violently persecuted Christians, he suddenly saw a bright light in the sky on his way to Damascus, fell to the ground, heard the voice of Jesus and was blind for three days (Acts 9:1–18). After this conversion experience, he became a missionary for early Christianity with tremendous zeal (hyper-religiosity?) and wrote a multitude of writings (polygraphy?). The Prophet Mohamed (570–632 CE) has been cited by some authors as a further example of a combination of excessive religiosity, hypergraphia and tendency to violence in the context of temporal brain epilepsy. According to this, he repeatedly experienced intoxication-like brief changes of consciousness with falling down, visual and acoustic hallucinations similar to those of the apostle Paul [24]. A new analysis of the historical sources on which these assumptions are based, however, concludes that this diagnostic classification is based on errors in interpretation and translation of the original sources and that Mohamed did not suffer from epilepsy [28]. Due to similar symptomatology, the same suspected diagnosis was made for several other founders of religions and sects as well as a number of Catholic saints [24].

Psychotic illnesses, which are predominantly schizophrenic but can also be caused by drugs, can also lead to abnormal religious feelings and even religious delusions and violent outbreaks (see Chap. 10). In these psychoses, the main focus of brain dysfunction lies in the limbic structures of the temporal brain and thus in the same brain regions as in temporal brain epilepsy [29].

The abnormal expression of aggressiveness and religiosity in malfunctions of emotionally relevant limbic regions of the temporal brain suggests that these brain areas also play an essential role in the development of normal aggressiveness and religiosity. It also makes it understandable that the cerebral cortical areas in the frontal brain and temporal brain, which control the activity of the limbic temporal brain, are more active during religious feelings in functional brain imaging studies [22, 23].

The occurrence of both pathological religiosity and pathological aggressiveness in psychoses and temporal lobe epilepsy as a result of dysfunctions in the same brain regions is a further indication of the close proximity of those brain circuits that are responsible for violence and religion.

Limits of Cognition and Knowledge

When discussing brain-biological correlates of religious feelings, there is one aspect that should not be disregarded. It is indisputable from a brain-scientific point of view that the cognitive capacity of our brain is bound to the possibilities of its structure and function. We can no more transcend these limits of cognition set by our brain physiology, even with the use of the most elaborate auxiliary means and despite all the cultural achievements of humankind, than an animal can transcend the feelings and behaviors predetermined by its brain. What lies outside the neurobiologically predetermined cognitive possibilities of our brain remains hidden from us. If it enters our world of imagination, we can believe in it—or not.

References

1. Armstrong K (2014) Fields of blood: religion and the history of violence; Chap. 7. Random House, London
2. Orientierung-M (2011) Zehn Gebote auch im Islam? https://www.orientierung-m.de/zehn-gebote-auch-im-islam/
3. Elias AA (2016) Faith in Allah—the ten commandments in Islam. https://www.abuaminaelias.com/ten-commandments-in-islam/. Accessed April 16, 2021
4. Afsaruddin AJ (2013) Encyclopedia Britannica. https://www.britannica.com/topic/jihad
5. Gharaibeh M (2014) Wahhabism and Salafism Shared foundation—different method. Quantara. https://en.qantara.de/content/wahhabism-and-salafism-shared-foundation-different-methods
6. Christianity Today. Theodosius I—Emperor who made Christianity "the" Roman religion. https://www.christianitytoday.com/history/people/rulers/theodosius-i.html. Accessed April 20, 2021
7. Armstrong K (2014) Fields of blood: religion and the history of violence; Chap. 8. Random House, London
8. Meyer JR St. Bernard of Clairvaux. Encyclopedia Britannica. https://www.britannica.com/biography/Saint-Bernard-of-Clairvaux. Accessed April 20, 2021
9. Jones D (2017) The Templars: the rise and spectacular fall of god's holy warriors. Viking, New York
10. Armstrong K (2014) Fields of blood: religion and the history of violence; Chap. 9. Random House, London
11. Britannica E Massacre of St. Bartholomew's Day. https://www.britannica.com/event/Massacre-of-Saint-Bartholomews-Day. Accessed April 20, 2021
12. Adams S Siege of Drogheda. Encyclopedia Britannica. https://www.britannica.com/event/Siege-of-Drogheda. Accessed April 2, 2021
13. Armstrong K (2014) Fields of blood: religion and the history of violence; Chap. 4. Random House, London
14. Wikipedia.org. Testaments of the Twelve Patriarchs
15. Tikhonov V, Brekke T (eds) (2013) Buddhism and violence: militarism and Buddhism in modern Asia Vladimir. Routledge, New York
16. Victoria D (2006) Zen at war, 2nd edn. Lanham, MD, Rowman & Littlefield
17. Armstrong K (2014) Fields of blood: religion and the history of violence; Chap. 12. Random House, London
18. Wikipedia.org. Religious violence in Odisha. https://en.wikipedia.org/wiki/religious_violence_in_Odisha
19. Gerlach C (2010) Extremely violent societies: mass violence in the twentieth-century world. Cambridge University Press
20. Editors H (2010) co. Jonestown Massacre. A&E Television Networks. https://www.history.com/topics/crime/jonestown. Accessed April 21, 2021
21. Pletcher K (2021) Tokyo subway attack of 1995. Encyclopedia Britannica. https://www.britannica.com/event/Tokyo-subway-attack-of-1995. Accessed April 21, 2021
22. Harris S, Kaplan JT, Curiel A, Bookheimer SY, Iacoboni M, Cohen MS (2009) The Neural correlates of religious and nonreligious belief. Sporns O (ed). PLoS One 4(10):e7272. https://doi.org/10.1371/journal.pone.0007272
23. Kapogiannis D, Barbey AK, Su M, Zamboni G, Krueger F, Grafman J (2009) Cognitive and neural foundations of religious belief. Proc Natl Acad Sci 106(12):4876–4881. https://doi.org/10.1073/pnas.0811717106
24. Saver J, Rabin J (1997) The neural substrates of religious experience. J Neuropsychiatry Clin Neurosci;9(3):498–510. https://doi.org/10.1176/jnp.9.3.498
25. Ferguson MA, Nielsen JA, King JB et al (2018) Reward, salience, and attentional networks are activated by religious experience in devout Mormons. Soc Neurosci 13(1):104–116. https://doi.org/10.1080/17470919.2016.1257437

26. Tebartz van Elst LT, Woermann FG, Lemieux L, Thompson PJ, Trimble MR (2000) Affective aggression in patients with temporal lobe epilepsy: a quantitative MRI study of the amygdala. Brain 123(2):234–243. https://doi.org/10.1093/brain/123.2.234
27. Devinsky O, Lai G (2008) Spirituality and religion in epilepsy. Epilepsy Behav 12(4):636–643. https://doi.org/10.1016/j.yebeh.2007.11.011
28. Aziz H (2020) Did Prophet Mohammad (PBUH) have epilepsy? A neurological analysis. Epilepsy Behav 103:106654. https://doi.org/10.1016/j.yebeh.2019.106654
29. Bogerts B (1993) Recent advances in the neuropathology of schizophrenia. Schizophr Bull 19(2):431–445. https://doi.org/10.1093/schbul/19.2.431

Chapter 21
Conclusions for the Prediction and Prevention of Violence

A profound knowledge of the multilayered, interrelated causes of the different manifestations of violence is a prerequisite for the most accurate possible prediction and effective prevention. Prediction of violent events is only possible within narrow limits and is associated with a considerable prognostic uncertainty; preventive measures, on the other hand, can be successful.

Limits to the Predictability of Individual Violence

Given the diversity of manifestations and causes of violence, simple models for prediction or prevention cannot be expected. Rarely is there one single cause (e.g., delusional mental disorder) underlying acts of violence; much more frequently, several factors or a whole bundle of causes occur together, for example, hereditary disposition + early life experience + provocation + alcohol. The 12 partial causes shown in Fig. 21.1 result in an unmanageable number of possible combinations, which, moreover, are hardly calculable in the course of their coincidence.

Some partial factors such as genetic make-up or pre-existing brain structure defects cannot be changed anyway; however, they only predispose to violence if epigenetic influences are present (e.g., through early traumatization; see Chap. 5) or when violence-provoking social stress comes along. Predicting situations that trigger individual violence is just as imponderable as predicting the occurrence and course of mental disorders. Psychoses and mood disorders, which are associated with a slightly increased risk of individual violence, develop slowly in young adulthood without a recognizable cause and are usually not recognized as such by the outside world in the early stages. However, they can sometimes be recognized by early warning signs (see Chaps. 10 and 16).

Even in a previously healthy brain, the neuronal centers of violence can become active due to illness without external psychosocial causes being present. The stress

Fig. 21.1 Multifactorial causal structure of violence. Own illustration

threshold, which must be exceeded for violence to erupt, can drop so low due to pathological brain processes that minimal trigger situations are sufficient. As is the case with all other elementary emotions such as depressive mood, euphoria or fear, aggression also shows smooth transitions from the normal psychologically understandable forms to the pathological range. The causes of brain diseases that cause emotional control to fail, including for aggression and violence, can be manifold; scientific research into the causes is increasingly focusing on inflammation-like processes in emotionally relevant brain areas [1–3].

Legal experts are often asked for a prognosis on the future dangerousness of convicted offenders as an aid to judicial decision-making. In addition to a thorough assessment of the overall psychopathological and social situation, assessment scales are used for this purpose, the sum values of which can be used to formulate probabilities for future violent offenses. These crime prognosis scales include the Psychopathy Checklist (PCL-R) [4, 5], the HCR-20 Scale [6, 7] and the Violence Risk Appraisal Guide (VRAG-R) [8].

The Psychopathy Checklist (PCL-R) assesses the expression of personality traits such as "tricky, articulate phoney," "inflated self-esteem," "pathological lying," "deceitful manipulative behavior," "lack of remorse or guilt," "lack of empathy," "impulsivity," "early behavioral problems" and "antisocial behavior."

The HCR-20 scale assesses aspects of past biography (H = history), including violent tendencies, level of psychopathy and psychological symptoms (C = clinical items), and social situation and future prospects (R = risk).

The Violence Risk Appraisal Guide (VRAG-R) assesses, similar to the HCR-20 scale, early family situation, school problems, alcohol and drug problems, partnership status, criminal history, age, conduct disorder and antisociality.

Using the sum values of these scales, risk categories for recidivism probabilities can be created. The predictability of violence derived from these scales is limited. In the VRAG-R, a recidivism probability of 76 per cent was determined for the highest risk category for a five-year period. This means that a quarter of the violent offenders who are assumed to be at the highest risk will not become violent again during this period [9]. This is true for offenders who are on trial for offenses they have already committed. The prediction of first-time violent offenses by previously non-offending suspects is even more uncertain.

Just as with the mid-term or long-term prediction of the weather or stock market courses, an exact forecast of the occurrence and course of all multifactorial systems is hardly possible; neither is it for individual acts of violence. Therefore, we will continue to be surprised by such events in the future, either through the media or through personal experience.

Predictability of Collective Violence

Group violence, be it by hooligans, left-wing or right-wing radical groups, "autonomists," including associated riot tourists, and at its extreme, pogroms, genocide and wars, is somewhat more predictable than individual violence. Group violence rarely occurs spontaneously, but first requires an underlying social situation, a group organization, the development of communication structures and execution planning, which takes time and effort. In addition, there must be an image of the enemy and an ideology by which violence is justified. Riots at football events or political demonstrations can be predicted with a certain degree of probability by those familiar with the scene. Terrorist groups, however, naturally plan their actions in secret, and in this context some Internet forums are becoming increasingly important. The prevention of terrorist attacks depends less on socio-political measures than on the drainage of financial sources and the efficiency of intelligence surveillance possibilities, which, however, are sometimes in conflict with data protection requirements.

The ability to influence political or social situations that promote violence (e.g., deficient monopoly on the use of force or disintegration of marginalized groups) often reaches its limits, as does the repression of violent ideologies, obsession with power or hedonistic group violence by hooligans or other riot squads.

Those who have once experienced the effect of collective group violence as uplifting the self-esteem or even as intoxicating will take the next opportunity to do so again. Similar to other addictive phenomena, once the brain's reward system has been activated by the perception or execution of violence, it will mandate the other brain functional circuits to seek out the same or similar situations again. Individuals who are unable to operate their reward system through social successes of a different kind will be particularly vulnerable to this.

Phylogenetic Disposition to Individual and Collective Violence Remains Unchanged

Moreover, the phylogenetic disposition to individual, collective, reactive and proactive violence resulting from hundreds of thousands of years of human development is and remains present in each of us as a negative legacy of human development. This will not change in future generations. In any case, the brain-biological preconditions for this are firmly anchored in every human being, even if many are not aware of them and for the vast majority of people they are pushed far into the background by prosocial attitudes. The readiness to overvalue one's own group, striving for dominance, devaluing others, resulting in exclusion and even group hatred and collective aggression, which can be derived from the phylogenetic development of *Homo sapiens*, will also remain in the world. Due to our predisposition to this, it will spontaneously arise again and again in groups of people and entire societies if it is not stopped by intensive cultural efforts including education and upbringing with the teaching of prosocial values.

Otherwise, it is to be feared that not much will change in the course of human history, which from prehistoric times to the present has been characterized by violence, sometimes more, sometimes less intensively, with phasic fluctuations. In no way can it be ruled out that the developments of periods of war and peace in the future will be similar to those of the past. There are many historical examples of how, both in Europe and in Asia, prolonged periods of warlike atrocities were followed by long periods of peace, which, however, were ended by new warlike violence or pogroms as soon as history was forgotten. The resurgence of right-wing populism 75 years after the end of the Second World War, combined with xenophobia, ethnic consciousness and national arrogance, can be interpreted as the first precursor of such a development, especially since in some countries rhetorically gifted populists with this kind of personality have become electable again.

Current Situation for the Prevention of Violence

In 2017, the following most important risk factors for violence were named by the Commonwealth Secretariat as targets for preventive measures [10]:

- Poverty, economic decline and unequal income distribution
- Displacement and movement of people
- Judicial or political corruption

- Adversity during childhood
- Challenges of personal and social identity
- Association with delinquent peers
- Social isolation.

At present, the assessment of these risk factors shows for many countries of the world a very different starting situation for the prevention of violence. According to the Global Peace Index 2020 [11], which summarizes the frequency of murder, manslaughter, bodily harm, use of weapons, public unrest and police presence in an index value, Europe remains the most peaceful region in the world, while the Middle East and North Africa region along with some Central American and African states remain the world's most dangerous regions. The most important prevention measures in these world regions must be political in nature, which means establishing a functioning administration including a stable state monopoly on the use of force, improvement of economic performance and education, fair distribution of resources, acceptance of the legal system, freedom of the press and low incidence of corruption. In other words, the most effective prevention of violent crime is provided by social systems with high standards of living, functioning education and legal systems, poverty reduction and appropriate integration of marginalized groups. But also in European countries, an intensified prevention is needed, especially with regard to the increase in right-wing extremist violence.

Prevention Principles and Projects

Prevention is most effective in childhood and adolescence because the brain is much more plastic in this phase of life than in adulthood; that is, the child and adolescent brain reacts faster and more sustainably with an adaptation of brain structure and function to psychosocial influences.

In 2010, the World Health Organization formulated seven important fields of work for the prevention of violence [12]:

- Development of safe, stable and nurturing relationships between children and their parents and caregivers
- Developing life skills in children and adolescents
- Reducing the availability and harmful use of alcohol
- Reducing access to guns, knives and pesticides
- Promoting gender equality to prevent violence against women
- Changing cultural and social norms that support violence
- Reducing violence through victim identification, care and support programs.

For school age, different prevention strategies have been developed that either include all pupils in a class or community (universal prevention) or only selected groups with increased risk (selective prevention, e.g., for those from a violent milieu) or only high-risk children or young people who have already attracted attention through problem behavior (indicated prevention) [13, 14]. Problem areas for such projects are a risk of stigmatization for the participants, difficult tangible effectiveness and often a lack of motivation to participate.

A particular challenge is the realistic assessment of the risk of future violent acts by adolescents or young adults. Special assessment criteria and scales have been developed for this purpose (e.g. VRAG-R, [8] NETWASS [15–17], RADAR-iTE [18], TARGET [19]), but their scores—similar to the prognosis scales developed for adult perpetrators of violence—only indicate degrees of probability; they cannot predict with certainty whether the person assessed will become violent or not within a foreseeable period of time. This can be explained by the fact that violent acts, even among juveniles, are usually caused by a concurrence of several causal and triggering factors (see Figs. 12.5 and 21.1), whose situational dynamics and future course are difficult to calculate. However, the risk of violence determined by the sum value of such scales in the form of a statistical probability level can provide valuable decision-making aids as to whether and in what form a young person should be integrated into a selective or indicated prevention program.

Special attention has to be paid to early developments tackling extremist tendencies and radicalization processes. The main strategy is to contribute to deradicalization by building new friendships, life planning, integration into work and awakening new interests.

Successful Evidence-Based Projects

Not all prevention projects have been evaluated positively so far, despite the strong motivation of those carrying them out.

A very comprehensive and up-to-date report providing an overview of successful evidence-based international violence prevention programs has recently been published [14]. This report describes 26 prevention strategies, the ways in which they are delivered, their effectiveness and the factors that influence their performance. Among others, the following prevention strategies are listed:

- Home-visiting programs for at-risk parents including teenage parents, ethnic minorities without a social network, or parents with substance abuse
- Child maltreatment programs
- Preschool learning and support programs
- School and classroom management to promote a safer school environment
- After-school programs
- Anti-bullying programs
- School programs for the prevention of child sexual abuse
- Cognitive behavioral programs for offenders
- Programs against violence in cyberspace
- Restorative justice programs for young and adult offenders
- Situational crime prevention in public areas.

A multitude of studies concludes that all of the strategies listed in this report are effective and lead to a significant reduction of violence [14].

It should be mentioned, however, that the effectiveness of these prevention programs depends not only on the underlying theoretical concept, the motivation and support for the social workers involved, but also on the surrounding social and political environment, educational level, ethnicities, cultural and socioeconomic factors. Thus, prevention programs that are effective in one region of the world might not necessarily be as successful in another [14]. Moreover, the various individual programs have to be supplemented by more general social development strategies [20].

The success of such programs also depends on sufficient personnel and financial resources as well as an intensive and smooth cooperation of the sectors of public life that are mainly responsible for this: the public health sector, the justice sector, the education sector, and the social service sector [21].

The question arises as to whether it is also possible to prevent violent incidents on the largest scale, sometimes of apocalyptic dimensions, such as state terrorism, pogroms, wars and genocide. The UN organizations created for this purpose in the post-war period, in particular the Convention on the Prevention and Punishment of the Crime of Genocide, adopted in 1948, have proved effective in many conflict situations, but by no means always.

The least susceptible to warfare, state terror and genocide are states with a democratically elected government, a multi-party system that guarantees plurality of opinions and a parliamentary-controlled monopoly on the use of force [22].

Final Remarks

Despite all our high technology and cultural achievements, our basic mental equipment anchored in the reptilian part of our brain is still at a Stone Age level. This can only be corrected through education and once again through education, especially in the form of conveying prosocial attitudes and moral norms.

The best and most sustainable prevention of violence is an education that, in the most malleable phase of life, that is infancy, imparts a stable bonding ability, interpersonal warmth and a sense of foreign values through intensive positive emotional care. If this prerequisite is not given, the effectiveness of social measures to prevent violence remains limited.

References

1. Goldsmith DR, Rapaport MH, Miller BJ (2016) A meta-analysis of blood cytokine network alterations in psychiatric patients: comparisons between schizophrenia, bipolar disorder and depression. Mol Psychiatry 21(12):1696–1709. https://doi.org/10.1038/mp.2016.3
2. Al-Diwani AAJ, Pollak TA, Irani SR, Lennox BR (2017) Psychosis: an autoimmune disease? Immunology 152(3):388–401. https://doi.org/10.1111/imm.12795
3. Bogerts B, Winopal D, Schwarz S et al (2017) Evidence of neuroinflammation in subgroups of schizophrenia and mood disorder patients: A semiquantitative postmortem study of CD3 and CD20 immunoreactive lymphocytes in several brain regions. Neurol Psychiatry Brain Res 23:2–9. https://doi.org/10.1016/j.npbr.2016.11.001
4. Hare RD, Clark D, Grann M, Thornton D (2000) Psychopathy and the predictive validity of the PCL-R: an international perspective. Behav Sci Law 18(5):623–645. http://www.ncbi.nlm.nih.gov/pubmed/11113965
5. Hare R (1991) The Hare psychopathy checklist—revised (PCL-R), multi-health systems. Toronto, Ontario
6. Douglas KS, Reeves KA (2010) Historical-clinical-risk management-20 (HCR-20) violence risk assessment scheme: rationale, application, and empirical overview. In: Otto RK, Douglas KS (eds) International perspectives on forensic mental health. Handbook of violence risk assessment, pp 147–185. https://psycnet.apa.org/record/2008-14279-008
7. Criminal Justice>Forensic Psychology>HCR-20 for Violence Risk Assessment HCR-20 for Violence Risk Assessment. Criminal Justice>Forensic Psychology. http://criminal-justice.iresearchnet.com/forensic-psychology/hcr-20-for-violence-risk-assessment/#google_vignette. Accessed 3 May 2021
8. Quinsey VL (2019) Violence risk appraisal guide (VRAG) and the violence risk appraisal guide-Revised (VRAG-R). In: The SAGE encyclopedia of criminal psychology. https://doi.org/10.4135/9781483392240.n531
9. Singh JP, Grann M, Fazel S (2011) A comparitive study of violence risk assessment tools. A systematic review and metaregression analysis of 68 studies involving 25,980 participants. Clin Psychol Rev 31(3):499–513. https://doi.org/10.1016/j.cpr.2010.11.009
10. Bellis M, Hardcastle K, Hughes K, Wood S, Nurse J (2017) Preventing violence, promoting peace—a policy toolkit for preventing interpersonal, collective and extremist violence. Commonw Secr Public Heal Wales
11. Institute for Economics & Peace (2020) Global peace index 2020: Measuring peace in a complex world. https://www.visionofhumanity.org/maps/. Published 2020. Accessed 2 Dec 2020
12. WHO(2010) Violence prevention: the evidence. http://apps.who.int/iris/bitstream/handle/10665/77936/9789241500845_eng.pdf?sequence=1. Published 2010. Accessed 5 May 2021
13. Gordon RS, An operational classification of disease prevention. Public Health Rep 98(2):107–109. http://www.ncbi.nlm.nih.gov/pubmed/6856733
14. Averdijk M, Eisner M, Luciano EC, Valdebenito S, Ingrid O(2020) Effective violence prevention, an overview of the international evidence. Cambridge. https://www.vrc.crim.cam.ac.uk/files/vp_report_-_digital_version_-_not_for_printing.pdf
15. Leuschner V, Fiedler N, Schultze M et al (2017) Prevention of targeted school violence by responding to students' psychosocial crises: the NETWASS Program. Child Dev 88(1):68–82. https://doi.org/10.1111/cdev.12690
16. Leuschner V, Bondü R, Schroer-Hippel M et al (2011) Prevention of homicidal violence in schools in Germany: The Berlin Leaking Project and the Networks Against School Shootings Project (NETWASS). New Dir Youth Dev 2011(129):61–78. https://doi.org/10.1002/yd.387
17. Fiedler N, Sommer F, Leuschner V, Scheithauer H (2019) Student crisis prevention in schools: the NETWorks Against School Schootings Programm (NETWASS)—an approach suitable for the prevention of violent extremism? Int J Dev Sci 13:109–122. https://www.researchgate.net/publication/338808615_Student_Crisis_Prevention_in_Schools_The_NETWorks_Against_School_Shootings_Program_NETWASS_-_An_Approach_Suitable_for_the_Prevention_of_Violent_Extremism

18. Bundeskriminalamt (2017) Presseinformation: Neues Instrument zur Risikobewertung von potentiellen Gewaltstraftätern [Federal Criminal Police Office. Press release: new instrument for risk assessment of potential violent offenders]. https://www.bka.de/DE/Presse/Listenseite_Pressemitteilungen/2017/Presse2017/170202_Radar.html. Published 2017. Accessed 24 Sept 2020
19. Ahlig N, Göbel K, Allwin M, Fiedler N, Leuschner V, Scheithauer H (2020) Testing for reliability of the TARGET threat analysis instrument (TTAI): an interdisciplinary instrument for the analysis of school shooting threats. In: Akhgar B, Wells, D, Blanco J (ed) Investigating radicalization trends: case studies in Europe and Asia, vol 1. Springer Nature Switzerland AG, Cham, pp 81–100
20. Haggerty KP, McCowan KJ (2018) Using the social development strategy to unleash the power of prevention. J Soc Social Work Res 9(4):741–763. https://doi.org/10.1086/700274
21. WHO (2020) Violence Prevention through multisectoral collaboration—an international version of the collaboration multiplier tool to prevent interpersonal violence
22. Rummel R (1997) Statistics of Democide: genocide and mass murder since 1900. Transaction Publishers, Rutgers University, New Brunswick, NJ

Index

A
Addiction, 11, 32, 71, 72, 79, 83, 84, 86, 89
Aggression center, 37, 39, 41, 44–47, 51, 79, 148, 149
Alcohol effects, 85
Alcohol in Europe, 84
Amygdala, 38, 41–44, 46, 52, 54–56, 63, 68, 75–77, 79, 85, 108, 130, 148, 168, 185, 205
Anomie, 120–122, 177
Antidepressants, 63, 74, 130
Antisocial personality, 51–53, 75, 76, 86, 130, 162
Appetitive aggression, 44, 106–108, 110, 118
Archaeological evidence, 21
Attention Deficit Hyperactivity Disorder (ADHD), 74, 75, 130, 132, 142, 143

B
Banality of evil, 93, 95
Behaviour in war, 182
Bertelsmann Transformation Index, 115, 117
Borderline, 63, 74, 76, 79, 97, 139
Brain biological correlates, 68, 108, 184
Brain diseases, 57, 127, 204, 205, 210
Brain injury, 75
Brain maturation, 128
Brain pathology, 51, 56, 57
Brain stem, 37, 38, 40, 42, 51, 62, 123
Brain stimulation, 38, 108
Buddhism, 195, 201

C
Cannabis, 85
Character of spree killers, 143
Character traits, 18, 29, 31, 34, 47, 52, 68, 77, 97, 107, 132, 178
Childhood aggression, 129
Childhood experiences, 30, 33, 83
Child maltreatment, 11, 214
Child sexual abuse, 190, 214
Cholerics, 78, 79, 97
Christianity, 195, 197–199, 205
Collective violence, 3, 4, 7, 8, 13, 67, 93, 105, 106, 114, 173–176, 181, 184, 203, 211, 212
Consumption of media violence, 131
Cortisol, 61–63
Costs of violence, 11
Crusades, 103, 198, 203

D
Darwinism, 17
Depression, 11, 71, 72, 74, 86, 119, 138, 141–143, 148, 165
DNA, 24, 29, 30
Drive theory, 92, 95
Drug effects, 74
Drug terror, 86
Durkheim, 120, 121

E
Economic prosperity, 119
Empathy, 9, 20, 25, 32, 34, 46, 47, 77, 96–98, 103, 110, 131, 132, 145, 168, 191, 205, 210

Environmental influences, 18, 29–31, 34
Epigenetics, 30, 35, 53, 209
Evolutionary psychology, 107
Exodus, 200

F
Family environment, 31, 34, 129, 132
Family studies, 31, 34
Fanatics, 78, 97
Female aggression, 68
Female violence, 67, 68, 138
Frame of reference, 183, 192
Frequency, 7, 72, 76, 78, 79, 113, 117, 122, 128, 190, 213
Frequency of violence, 84, 113, 114
Frontal brain, 43, 46–48, 52, 53, 57, 63, 68, 75, 77, 79, 85, 104, 127, 128, 130, 148, 168, 204–206
Frustration theory, 92, 95
Future risk, 150

G
Gender differences, 62, 67, 68, 129
Gene expression, 29
General Aggression Model, 95, 96, 191
Genes, 1, 17–19, 21, 25, 29–35, 53, 62, 67, 68, 92, 107, 129, 192
Genocides, 1, 3, 13, 105, 106, 160, 173, 175–178, 192, 202, 211, 215
Group aggression, 20, 101, 107, 173, 174, 184, 185
Group terrorism, 169

H
Hallucinogens, 85, 86
Hedonistic violence, 4, 103, 106–108, 110, 190, 191, 196, 203, 205
Hereditary factors, 31, 129
Heritability, 29, 31, 32, 34
Hinduism, 195, 201
Hippocampus, 41–43, 52, 54, 63, 130, 168, 205
Historical background, 156
Historical comparison, 13
Historical dimensions, 173, 175
Hobbes, 89, 90, 117
Homicide rates, 7, 8, 11, 25, 114, 119, 121
Hooligans, 1, 3, 9, 101, 106–108, 118, 182, 184, 211
Hormons, 25, 34, 46, 61–63, 79, 102, 182
5-HTT gene, 33, 34

Human evolution, 17, 174, 204
Hypothalamus, 37, 39, 41, 43–46, 62, 68, 108, 148, 189

I
I-cubed-model, 96
Imaging techniques, 51, 68, 89, 168
Indigenous peoples, 23, 177, 199
Indigenous populations, 24
Individual violence, 3, 8, 173, 209, 211
Information processing, 44, 45
Inquisition, 95, 103, 196, 199, 203
Intergroup conflicts, 25
Internet, 138, 140, 156–158, 162, 165, 167, 190, 197, 211
Islam, 157, 161–163, 195–197

J
Jihad, 163, 197
Judaism, 200

K
Killer games, 101, 106
Koran, 157, 162, 197, 200, 201

L
Leaking, 146, 147
Limbic system, 40, 42–44, 47, 48, 51, 52, 54, 62, 79, 85, 104, 108, 109, 123, 167, 185
Lombroso, 90, 91
Lone-wolf terrorists, 165

M
Machiavellianism, 97
Male violence, 10, 67, 119, 168, 182, 189
Mania, 148, 149, 205
MAO-A gene, 33, 34
Martial arts, 101, 102, 184
Mass psychology, 173, 184
Mental disorders, 1, 4, 30, 71, 74, 78, 79, 83, 127, 130, 131, 139, 143, 159, 160, 162, 165, 209
Middle Ages, 13, 20, 95, 103–105, 113, 176, 177, 198, 200
Milgram experiment, 94, 95

N
Narcissm, 78, 97

Natural selection, 18, 19
Neighborhoods, 104, 119, 120, 122, 180
Neocortex, 40, 42, 43, 46, 108, 109, 185
Neurobiological correlates, 47
Neuroimmunological causes, 148
Neurotransmitters, 34, 79, 85, 109, 130
New Testament, 197
Nucleus accumbens, 44–46, 85, 108–110, 185, 205

O
Old Testament, 174, 200
Oxytocin, 25, 34, 62

P
Partnership violence, 9, 10
Perpetrators' character, 9
Personality disorders, 30, 72, 74, 76–79, 83, 98, 141, 143, 144, 164, 166, 191
Phineas Gage, 53
Phylogenesis, 18–20, 24, 25, 35, 43, 92, 106, 107, 173, 203, 204
Phylogenetic aspects, 192
Phylogenetic disposition, 212
Pogroms, 3, 20, 184, 201, 202, 211, 212, 215
Police violence, 118
Post-Traumatic Stress Disorder (PTSD), 11, 31, 75, 76, 105, 165
Predictability of violence, 131, 209, 211
Prediction of violence, 34, 133, 209, 211
Pregnancy, 130, 189
Prevention, 3, 145, 146, 150, 209, 211–215
Prevention principles, 213
Primates, 20, 24, 25, 43
Proactive violence, 4, 168, 196, 212
Prognosis scales, 210, 214
Prosocial behaviour, 24, 25, 34, 45, 47, 48, 62, 110, 178, 204
Psychology of Terror, 159
Psychology of violence, 89
Psychopaths, 45, 51–54, 77, 97
Psychopathy, 32, 51–53, 77, 79, 97, 98, 132, 210
Psychostimulants, 86
Psychosyndromes, 52, 75, 78, 97, 148, 161

R
Rampage events, 147
Reactive violence, 4, 51, 108
Red Army Faction, 56, 156, 159, 167

Refugees, 12, 181
Religious believes, 204
Religious violence, 203–206
Reptilian brain, 42, 43, 108–110, 114, 190
Revenge, 4, 104, 105, 107, 115, 129, 140, 141, 143, 144, 146–148, 157, 161, 168, 196, 201
Reward center, 40, 44, 45, 85, 109, 185
Reward system, 44, 47, 48, 61, 68, 85, 105, 108–110, 191, 205, 211, 212
Right-wing extremism, 169, 181
Risk factors, 9, 76, 119, 123, 129, 131, 132, 146, 165, 177, 178, 191, 212, 213
Rousseau, 89–91, 93

S
Sadism, 4, 97, 98, 103, 104, 107, 109, 203
Salafism, 163, 197
Salafists, 156, 163, 165, 166, 197
Schizophrenia, 71–73, 85, 139, 143, 149, 166
School shootings, 140–142, 145–147, 150, 155
Sects, 202, 206
Selective breeding, 18, 19
Serial killers, 104
Serotonin, 33, 34, 61, 63, 74, 85, 130
Sexually dimorphic nucleus, 68
Social disintegration, 120, 122
Social environment, 18, 29–31, 33, 37, 43, 52, 72, 95, 113, 123, 131, 137, 159
Social identity theory, 179
Societal attitudes, 119
Spree killers' brains, 147, 148
Stanford Prison experiment, 94, 95
State monopoly on force use, 105, 114, 115, 117, 119, 121, 176, 213
State terror, 4, 175–177, 215
State violence, 4, 115
Stress, 11, 30, 58, 62, 63, 75, 95, 96, 115, 118, 139, 168, 209
Structural violence, 3–5, 117
Successful projects, 214

T
Tajfel, 178–180
Templar knights, 198
Terror attacks, 8, 155–158, 163, 165, 167, 211
Terror definition, 155
Terrorists' brains, 167
Testosterone, 61, 62, 67, 68, 102

Thucydides, 89, 90, 175
Torture, 4, 95, 103, 161, 176, 196, 202, 203
Toxic influences, 130, 132
Traumatization, 83, 129, 149, 163, 209
Twin studies, 31, 32, 34

U
Ulrike Meinhof, 56, 167

V
Violence against women, 10, 119, 190, 213
Violence fascination, 101, 106
Violence frequency, 72, 76, 84, 113, 114, 190

Violence history, 201
Violence risk, 210, 211
Violence statistics, 11
Violent offenders, 41, 51, 52, 57, 61, 71, 72, 75, 79, 84, 130, 144, 147, 148, 168, 191, 211

W
Wagner, Ernst, 54, 55, 148
Warning symptoms, 146, 147
Whitman, Charles, 54, 148

X
Xenophobia, 13, 156, 160, 181, 185, 212

GPSR Compliance

The European Union's (EU) General Product Safety Regulation (GPSR) is a set of rules that requires consumer products to be safe and our obligations to ensure this.

If you have any concerns about our products, you can contact us on

ProductSafety@springernature.com

In case Publisher is established outside the EU, the EU authorized representative is:

Springer Nature Customer Service Center GmbH
Europaplatz 3
69115 Heidelberg, Germany

www.ingramcontent.com/pod-product-compliance
Lightning Source LLC
LaVergne TN
LVHW010340260326
834688LV00036B/797